A MANAGEMENT GUIDE TO RETROFITTING WASTEWATER TREATMENT PLANTS

HOW TO ORDER THIS BOOK

BY PHONE: 800-233-9936 or 717-291-5609, 8AM–5PM Eastern Time

BY FAX: 717-295-4538

BY MAIL: Order Department
Technomic Publishing Company, Inc.
851 New Holland Avenue, Box 3535
Lancaster, PA 17604, U.S.A.

BY CREDIT CARD: American Express, VISA, MasterCard

BY WWW SITE: http://www.techpub.com

A Management Guide to RETROFITTING WASTEWATER TREATMENT PLANTS

Lawrence E. Quick
Superintendent
Robinson Wastewater Treatment Facility
Robinson, Illinois

LANCASTER · BASEL

A Management Guide to Retrofitting Wastewater Treatment Plants

a **TECHNOMIC**® publication

Published in the Western Hemisphere by
Technomic Publishing Company, Inc.
851 New Holland Avenue, Box 3535
Lancaster, Pennsylvania 17604 U.S.A.

Distributed in the Rest of the World by
Technomic Publishing AG
Missionsstrasse 44
CH-4055 Basel, Switzerland

Printed in the United States of America

10 9 8 7 6 5 4 3 2 1

Main entry under title:
A Management Guide to Retrofitting Wastewater Treatment Plants

A Technomic Publishing Company book
Bibliography: p.
Includes index p. 113

Library of Congress Catalog Card No. 97-61845
ISBN No. 1-56676-594-3

Table of Contents

Preface

This book is a handy reference on making the task of upgrading or retrofitting wastewater process equipment an easier and less costly job. Also included are tips on selling your ideas to your superiors, pre- and postproject activities, and certain management techniques I have found to be successful. These are actual examples from my own work, shown as Exhibits throughout the book. Also note that I have included an appendix titled Frequently Asked Questions. This book represents information that is believed to be accurate. However, the author and the publisher cannot assume responsibility for the validity of all materials or for the consequences of their use.

Because federal funding is scarce for massive upgrades and/or complete new WWTP construction, treatment plant operators, superintendents, managers, city councils, boards, etc. must get more creative on funding and coordinating process equipment replacements. Herein you will find hints, tactics, and procedures aimed at getting the "biggest bang for your public buck." During the 1970s and 1980s, through grants, the federal government paid 80% of costs to build new or expanded wastewater treatment plants, pumping stations, and collection system renovations. The majority of the grants were to upgrade primary treatment facilities to secondary, and secondary to tertiary treatment status based on Clean Water Act regulations. If your facility was fortunate enough to receive grants, you were in good shape for approximately 20 to 30 years (depending on community growth rates). Because most wastewater treatment facilities are designed to last 20 years, many of the new or expanded facilities in the 1970s and 1980s are reaching the end of their service life. Some may have reached it sooner due to growth beyond the expected rate, inadequate preventive maintenance, or design inadequacies when built. Now you have identified problems with insufficient aeration capacity, equipment mechanical failure, insufficient pump station capacity, infrastructure deterioration, etc. and need to do something about it before you violate your NPDES permit (if you haven't already). This equipment seems very costly to replace because you now must pay 100% compared with 20% with the grants.

This book should prove helpful to those who find themselves in this situation and who need assistance on resolving facility problems. I have made an effort to put the text of this book in common operator language to allow easier understanding.

LAWRENCE E. QUICK
Superintendent
Robinson, Illinois WWTF

CHAPTER 1

Introduction

Many wastewater treatment plants (WWTP) are in need of replacement and/or upgraded equipment. The equipment itself is typically about 25 to 50% of the total project cost, which cannot be changed much. However, the remaining 50 to 75% (engineering, installation labor costs, and project management) may be whittled down, depending on how active and creative the project coordinator (yourself) wants to be in the process. When EPA funded 80% of project costs in prior years, it was no big deal to have an "open pocketbook" attitude. Those days have changed forever, and so have procurement procedures for projects.

RESOURCE ORGANIZATIONS

Throughout this book will be references pertaining to solicitation of professional help and/or resources. Please take these references at face value. Nobody knows more about what you or your staff is capable of achieving than yourself. For example, remember the last time you "farmed out" a project, and while you observed the project being performed, you thought to yourself "we could do that." Then you probably could. You may need to challenge yourself and staff to achieve the results, but, typically, if you work well as a team, you will rise to the occasion. And by the way, there is no better feeling and morale booster than to achieve a goal by performing tasks in-house. I have always viewed teamwork as the most important variable in achieving goals and "personnel" as the most valuable asset I have. Always keep your team informed about your plans and ask them for input and ideas. They may come up with ideas that will make projects easier or less costly. Also, utilization of their ideas will create an atmosphere of willingness to both help accomplish the tasks as well as take ownership in the results. If you do take on tasks in-house, be aware of unplanned problems. If this happens, don't be afraid to ask for help. Because information is power, all you may need is a boost.

Remember that you are never alone and rarely face an issue others have not faced in the past. If there are questions, do not be so stubborn as to not ask for help from others. Help may come in the form of other published text (magazines, manuals, EPA guidance manuals, etc.), neighboring WWTP operators or operator groups, etc. Some of my favorite resources are operators groups and associations, i.e., Water Environment Federation, Illinois Association of Water Pollution Control Operators, Illinois Water Environment Association, and Illinois Association of Wastewater Agencies. Many different states have their own groups and associations. Not only do the groups offer specialized training by means of workshops and conferences, they also offer books, magazines, and flyers that can keep you updated on current events, topics, and issues.

COMPUTER PROGRAMS

Another very important resource is the control authorities (IEPA, USEPA, etc.). These individuals and groups are always willing to lend a hand to help your facility. They have a vast amount of literature, computer programs, manuals of practice, and data that you can usually obtain for free or at a minimal cost. For instance, IEPA has provided me with a computer program geared for reporting (Yearly Pretreatment Program Report "PRETREP"). USEPA has provided me with computer programs as follows:

(1) "Model Pretreatment Ordinance" (This program contains all verbiage necessary to develop and design a pretreatment ordinance. Because the program is offered in Word Perfect format, you can modify the text to fit your specific application.)
(2) "PRELIM" (This program is designed to aid in the development of "local limits" contained in a Pretreatment Ordinance. You enter all sorts of facility design data, stream standards, sludge analysis data and criteria, activated sludge inhibition levels, etc., and the program will compute technically based limits for your facility.)
(3) The "BEN" Program (This program aids in the calculation of fines in the case of noncompliance from a significant industrial user.) Again the data are entered, and fines are computed based on the amount of money the industry saved by being noncompliant. All of the computer programs are user-friendly and are geared toward saving you time, money, and headaches.

COMMUNICATION OF RESULTS

One of the most crucial elements that is often overlooked by municipal operators is communication of results (Chapter 9 is entirely devoted to this subject). There are a few main reasons for this breakdown in communication. I believe most importantly is the fact that public officials in smaller, and even medium-sized communities, are "part-time" status. These individuals rely heavily on their operators to keep them updated on issues. If communication of problem issues is not addressed, then more than likely communication of success is as well.

Of vital importance is communication of facility achievements. Because wastewater treatment facilities and their operations are generally not the hot topic at the coffee shop (unless of course it is negative news), it is not hard to understand why exposure by the media is hard to come by. However, because everyone knows that energy (electricity, natural gas, etc.), chemicals (polymer, chlorine, sulfur dioxide, etc.) as well as upgrade and retrofit projects of any magnitude are expensive, exploitation of savings in these areas is vital. For instance, a savings of $30,000.00 a year on electricity, which may keep your city from raising sewer user rates, will definitely get attention. This may be obtainable by installing a more efficient aeration system (see Photos 1.1 and 1.2).

Also, a good rapport is necessary with local media (newspaper, radio stations, television, etc.) so that a phone call will get them interested in covering all positive events. Contact their reporters and maybe start out with smaller projects at first and build to larger ones. An important aspect to note is always give all the different media types a chance to report. I have found this promotes a competitive atmosphere among the media to get involved in covering facility issues. Do not get caught up in the scenario of imagining the public is aware of your positive results. They definitely are not, unless someone makes a big deal out of it. For instance, why do private operation and maintenance firms always have press releases on accomplishments? The reason is that they need the exposure to retain their contract in that community. Municipal operators need to view their job in the same exact manner.

Building trust and integrity is another very important task of municipal operators. Regardless of past experiences and facility problems, there is no better time than now to begin build-

Photo 1.1. Empty aeration system.

Photo 1.2. Fine bubble diffusers save energy by reduced horsepower on-line to achieve desired D.O. levels.

ing trust with your council people and the general public. The trust may already be there. If it is, build on it. It is no secret that contract operation and maintenance firms will do this and do it very well.

Public image is another important aspect. Municipal operators should stress not only to themselves but also to their entire crew the importance of their appearance in the public's eye. This can range anywhere from being very cordial to the public (regardless if you are right or wrong), exercising good ethics, maintaining facilities asthetics, being active in community groups, etc. For instance, let's say someone had a backed up sewer lateral, and the city has identified it is the property owner's responsibility because the main sewer is not plugged. Don't just blatantly say, "It is your problem; we're out of here." Whether they are right or wrong, hear them out. You could maybe give them some ideas on how they could fix their problem. Remember, they are the consumer, and the municipality is the seller of a service. It is surprising how effective just listening to a disgruntled citizen can be. They want to be heard and feel like they are being treated fairly. Typically, they will come around and understand. If they are not listened to, you have lost a supporter. If they are listened to, you have gained a supporter. This support may build slowly over time, but eventually the word will get around that "those wastewater guys are OK."

CHAPTER 2

Status of Existing Facilities

Although certain equipment may have an obvious need for replacement, others may not be so straightforward. The facilities operation and maintenance staff (may be the same individuals) should first get together and discuss existing equipment. Because the goal of a water pollution control professional is to meet or exceed your NPDES permit requirements, equipment or infrastructure first to be scrutinized should be critical items necessary for NPDES compliance. If immediate problems are with chlorination equipment, then efforts should be focused there, not on smoke testing the entire sewage collection system. If sanitary sewers are overflowing into creeks due to collapsed pipes, roots, or inadequate pumping station capacities, address the collection system issues first. A good place to start if regulatory agency problems are an issue, is with their compliance inquiry letters, better known as inspection result letters. Keep focused on the "Big Picture." Although some of the issues may not seem as significant to you, they may be very important to the regulating agency. Please keep in mind that these individuals are typically eager to assist if asked. The better your facility is operating, the easier their job is.

IN-HOUSE EVALUATION

A comprehensive evaluation should be performed on all existing facility equipment. A separate data sheet should be filled out for each individual process. For instance, if you have a duplex pumping station, document past problems with packing or mechanical seals, bearings, impeller wear, etc. and the estimated costs associated with the problems. Also, estimate the amount of man-hours you or your staff have spent in the past year on corrective maintenance to keep the equipment in service. The age of the equipment, general appearance, nameplate data (motor horsepower, pumping capacities, 1 or 3 phase, etc.) should all be on the data sheet. (A sample equipment data sheet is contained as Figure 2.1.)

Design capacity is a major factor in identifying the need to upgrade or retrofit. For instance, an activated sludge reactor that was designed for 5000 lb/BOD/day and has been treating 6000 lb/BOD/day creates a problem. Boards and councils seem to understand the need for equipment upgrade much better when explained in terms of design capacity. The regulatory agencies are also very scrutinous when this scenario exists and nothing is being done to rectify the problem. If permit violations do not occur, you may be all right, but if discharge limits are exceeded, EPA may implement enforcement actions (fines and mandates) or place the system on Restricted Status (no new hookups to your collection system). Both of these scenarios are very unattractive, especially if the community wants continued growth. Conditions, such as these, will also get the immediate attention of the chamber of commerce and other community growth groups. It is impossible for them to market the community if the wastewater facilities are either at or above design capacity.

The information concerning facilities design capacities is typi-

NAME OF EQUIPMENT:______________________________.

ELECTRICAL MOTOR DATA:

MOTOR #1	#2
MOTOR #1 MFR:____________________.	#2 MFR:____________________.
PART#:____________________.	PART#: ____________________.
HP:____________.	HP:____________.
FRAME:__________.	FRAME:__________.
TYPE:__________.	TYPE:__________.
CODE:__________.	CODE:__________.
HZ:____________.	HZ:____________.
PHASE:__________.	PHASE:__________.
RPM:___________.	RPM:___________.
VOLTS:__________.	VOLTS:__________.
AMPS:__________.	AMPS:__________.
DUTY:__________.	DUTY:__________.
SF:____________.	SF:____________.
INSULATION CLASS.__________.	INSULATION CLASS:__________.
BEARING DRIVE END:__________.	BEARING DRIVE END:__________.

SIZE:____________________.
CAPACITY:________________.
SERIAL No:____________________.

EQUIPMENT
HISTORY:____
____
____
____
____.

EXISTING
OPERATING
STATUS:____
____
____
____.

PAST
REPAIRS/COSTS:____
____
____
____.

EXPECTED REMAINING SERVICE LIFE:__________________ YEARS.

COMPLETED BY:__________________________. DATE:____ /____/____.

Figure 2.1. Equipment identification record.

cally contained in NPDES permits and reference operation and maintenance manuals. The design capacities should be contained in these documents. If they are not, the original engineering firm that designed the facility may have them. The engineering firm may also be able to calculate the design capacities based on tank sizes, configurations, and aeration capacities if necessary. More sampling and testing may be required to identify actual WWTP influent flows versus current loadings. The best results will be obtained by using 24-hour composite samples. They will give a representative idea of the overall daily loadings. Three or four samples collected during an eight-hour shift and mixed together does not accurately represent actual loadings. If portable or stationary composite samplers are not available, you may be able to borrow or rent a couple from a local engineering firm or neighboring treatment plant.

Some equipment is designed for hydraulics (pumps, clarifiers, and sand filters), e.g., gal/ft^2/day, and some for organics (blowers, chemical feed equipment and tank volumes/configurations), e.g., lb/BOD5/day. If the facility still has the original operation and maintenance manuals, this information is relatively easy to obtain.

PROFESSIONAL EVALUATION

If you feel you do not have the time or expertise to perform the evaluation yourself, then other cost-effective alternatives are available. Choosing a private firm to perform the evaluation based on the following criteria is an option:

- Get a quote from a wastewater engineering firm to evaluate all facility equipment. Be specific by stating that you want an equipment evaluation and recommendations for restoration and/or upgrade. You do not need a complete comprehensive facilities plan!!! Typically, it is not required to bid engineering assistance. (Refer to Chapter 4 for more detailed information on selecting a firm.) Choose a firm that you feel comfortable with to perform whatever tasks are necessary. I have found that certain engineering firms are better in specific areas. Some may be better at building bridges, airports, water treatment plants, etc. If you do choose to use an engineer, pick one with a good wastewater track record. Ask for a statement of qualifications from each potential firm for your review.

- Publish a request for proposals from contract operations and maintenance companies to evaluate the facility. If such a firm is located in your area, it may be best to contact them solely to begin with. Be sure to specify in the request for proposals that the evaluation data shall be contained with their final bid. Although it may be a hassle if you choose not to contract with one of the firms, copies of their evaluations will be on hand for utilization if necessary.

Typically, it will contain an individual equipment breakdown, needed repairs or replacements, and an initial maintenance and repair budget. This information, collected and assimilated by evaluation professionals, will prove very valuable when presented to your superiors.

We are very fortunate to have a professional equipment restoration company in our area (Environmental Resources Inc., tel.: 800-691-0998) which specializes in water and wastewater treatment equipment retrofits and upgrades. They have the practical expertise and process equipment knowledge to help WWTP operators with the total project or with the parts that you do not feel comfortable with.

PRIORITIZING

You must now begin the process of prioritizing needed upgrades based on how critical they are to the facilities compliance status. Take into consideration the amount of money that can be budgeted for capital replacement. Of course, this is to be done, assuming adequate funds are not available to perform all upgrades at once.

Also, consider effects on upstream and downstream processes, i.e., it would be ridiculous to upgrade the influent pump station for greater capacity if clarifiers and sand filters could not handle the added hydraulic load.

During the complete process, be completely up front and honest with superiors (board, council, etc.). Make sure they understand the importance of continually upgrading the existing facilities. Invite them to the plant or collection system and show them first hand problems and issues the facility is facing. When explaining issues to them, stick with nonsubjective data (such as design life of the equipment or hydraulic and/or organic basis of design). The new equipment may save money and man-hours; don't forget to note this rationale. Remember, design data are very important so that your superiors can make an informed decision based on the facts.

Avoid the old descriptives: "It just doesn't work anymore," "It never worked right," "We just need to replace it," etc. If questioned on these types of descriptions, you may be at a loss for words. Use descriptions, such as "It is operating at 110% of design capacity," "It requires 50 man-hours a week to keep it running," "Repair costs have exceeded $20,000.00 this year," "It has reached the end of its design life," etc. The board members will also have these data to present to John Q. Public if questioned.

CHAPTER 3

Planning

Now that the problems have been identified, necessary upgrades have been prioritized, and the information has been conveyed to the board, it is time to put together the plans of action. The best way to do this is to start with a short-term plan (1 year), follow-up with a medium-range plan (5 years), and then move on to a long-range plan (10 to 20 years). The plans should be configured as listed below. Don't forget to include rolling stock (trucks, backhoes, riding mowers, etc.). Generally, any single item over $1,000.00 should be contained. Normal operating supplies should not be contained in the plans but reflected in the yearly budget.

SHORT-TERM PLAN

The items listed should be the most immediate concerns. For instance, if you have been close to violating your permit due to erratic chlorination or dechlorination equipment, this should be in the 1-year plan. If your aeration tanks have failing safety rails, this should be in the 1-year plan.

(1) List each requested upgrade based on its priority.

(2) Briefly explain the scope (description) of the proposed upgrade.
(3) Estimate the cost of the upgrade. Call equipment representatives, fellow WWTP operators, or an engineering firm to help with the estimates. You can estimate that equipment will be about 25 to 50% of the total cost.
(4) Put the plan in a neat format. Use a word processor or typewriter or hire a temporary secretarial service to type it professionally. It doesn't need to be super high tech, but the plan will be much more appealing if it looks nice (see sample contained as Exhibit 3.1). Present the plan to the board. Give them time to look it over and ask questions. Usually a week before a committee or council meeting is sufficient. Try to get them to buy into the plan. Show them you are committed to whatever is necessary to implement the plan. A very good selling point is performing any tasks possible with in-house labor, i.e., changing sand filter media, retrofitting pumps with original manufacturer parts, etc. This is where a good idea of the capabilities and time availability of your crew is necessary. Don't hesitate to challenge them, because they could really surprise you. Also, let the board know that the costs are estimates and actual costs could be a little higher or lower than what is projected.

MEDIUM-RANGE PLAN

The medium-range plan should be similar to the 1-year plan except that you are envisioning the future. Is there equipment at the facility that is not on the verge of failing but has indications of wear and tear? Is there a need for recommended spare parts that could take weeks to get for existing equipment, e.g., gear reducers, bearings, mechanical seals, impellers, etc.? This plan is a good place to document these types of needs. Medium-range plans may need to include larger projects, such as major capacity upgrades. They may include installation of a more efficient diffuser system, another clarifier, nutrient removal equipment, etc. (see sample contained as Exhibit 3.2). It may also be a good idea to list some of the rationale for all plans.

Let them know you are not "shooting from the hip." *Note:* Please be advised that by taking this initiative, you are putting yourself and your integrity on the line. That is why considerable thought must be placed on personal development of these plans. However, if you feel you are up to the task, money savings, morale boosting, pride in workmanship, and a general atmosphere of teamwork will be the result.

If you want to begin this process but are afraid to perform them all at once, start with a 1-year plan, work through and finish it, and then move on to a 5-year plan. If the plan is put together by yourself, let the board know that. Also, let them know that plans may need to be changed or modified in midstream due to certain unknown variables. This will be a disclaimer that will allow some flexibility due to unknown conditions, changes in regulations, etc.

LONG-RANGE PLANS: "CRYSTAL BALL PLANNING"

The development of long-range plans may require the solicitation of an engineering firm (see Chapter 4). Typically, these plans take into consideration historic growth of the community, current status, and anticipated growth in the future. Because long-range planning is basically guesswork, the need to be careful not to over, or undersize proposed upgrades is a concern. If the upgrades are oversized, the existing sewer users are liable for retiring any debt associated with major upgrades.

If they are undersized, they may not provide adequate capacity to allow new industry to connect without them pretreating their wastewater. I tend to believe that conservative is better than liberal in these instances. Credibility can be lost forever if sewer customers are left paying for added capacity that is never utilized.

CAPITAL REPLACEMENT BUDGETING

So the board has given approval to proceed with the plan(s). "Great!" Now what? Start with the most detrimental project(s). Do not start too many projects at the same time. This can become overwhelming to administer and create an environment in which all projects are "watered down." Choose the most important and concentrate on them. Don't forget to include the project(s) in the upcoming fiscal year budget (see sample budget contained as Exhibit 3.3). My rationale has always been to budget one or two larger projects and a few smaller projects a year. This tends to not overwhelm board members. Also, when the request is brought to the board for a vote, remind them, "This was in my plan (the plan they bought into) and budgeted for this year." This point is very important because board members tend to forget all of the projects that have been planned and budgeted for. Just remind them so they know the project was not an unplanned expense that may put you over budget. Try to estimate costs as close as possible because if you are relatively close, you will build trust for future budget requests.

EXHIBIT 3.1

City of Robinson Wastewater Treatment Facility
One-Year Facilities Plan

This plan includes all aspects of the Wastewater Treatment Facilities, Safety, Operations, Sludge Disposal, and Collection System. The items contained in this plan are listed by priority.

Safety

Since safety is our most important issue, I felt we needed to address it first. I am developing a safety manual that includes safety rules, procedures, and Emergency Action Plans. This will be a comprehensive plan which covers all potential hazards associated with the facilities. This plan will be completed by 7/01/94. A Confined Space Entry program is also being developed for the facilities. This plan will contain procedures, entry permit guidelines and safety equipment use. This plan will be complete by 7/1/94.

We will hold monthly safety meetings which will include videotapes, lectures, handouts, and a quiz. The safety meetings will make the employees aware of the hazards that are around them. These meetings will begin in May 1994.

Facility Operations

The Contact Stabilization Plants are operating well at this time but are at peak design organic loadings. There are a few upgrades that need to be made to the facility to give us a buffer zone to handle warmer temperature biological activity and higher loadings.

a. The existing Contact Stabilization Plants need to be retrofitted with a floor cover design fine bubble diffuser arrangement. This retrofit will give us an extra 1,000 lb BOD/day capacity if coupled with a new positive displacement blower.

   A new 100-hp positive displacement blower should be added to take the place of one existing 75-hp centrifugal blower. The existing centrifugal blowers tend to "surge" (a condition caused by excessive back pressure on the blowers discharge) in warmer weather.

   The blower discharge line which channels the air to the tanks has many leaks. These leaks make the system very inefficient. This line must be dug up and repaired.

   With the addition of these two retrofits and repair of the air lines, we should enjoy power savings, as well as increased plant capacity and treatment.

Estimated cost: $140,000.00

b. While the Contact Stabilization plants are emptied, we need to inspect them for damaged paint, remove any grit buildup, and make any adjustments or repairs that are necessary.

Estimated cost: $20,000.00

c. The RBC building needs to have the channels drained and cleaned. There is a buildup of sludge and grit in the bottom that makes them much less efficient.

Estimated cost: $2,000.00

d. The west tertiary clarifier needs the scraper drive gear reducer removed and inspected. It has a very noticeable vibration, which leads me to believe there are worn or damaged parts.

Estimated cost: $2,000.00

e. The Tertiary Sand Filters need to be put back on line. They are inoperable because of fouled media and drainage systems. These filters catch any solid matter that makes it through the previous treatment facilities and returns it back to the influent flow stream.

Estimated cost: $20,000.00

Sludge Disposal

The owner of the current permitted property which receives our sludge has informed me that we will not be allowed to apply sludge there anymore. We will try to get his approval to utilize this property until we get new agricultural land permitted.

a. We will secure a tract of land closer to Robinson to dispose of our sludge. The land application permit from IEPA needs to be modified to include this permitted property. All necessary documents are being prepared as of 5/10/94. We should have the land permitted by 10/01/94. This newly permitted site should be more cost-effective for the City considering the closer distances the sludge will need to be transported.

Estimated permit cost: $3,000.00

b. The method and frequency that sludge is land applied will be determined by the window of time the farmer will allow us to enter the property. Considering the storage area we have on site at the Plant, we should have ample room to allow us spring and fall sludge hauling events.

Stabilization Lagoon

a. We are in the process of lowering the lagoon level to a point where we can install the permitted emergency overflow pipe. This pipe will be used on an emergency basis. If the lagoon level should rise to a point of emergency overflow caused by excessive rainfall, this pipe will drain excess water to the creek opposed to over the lagoon berm. The lagoon procedures will be documented in the form of a management plan in the facility Standard Operating Procedures Manual.

Lift Stations

I have not had time to get to all lift stations yet. The lift stations I have been to look as though they need a thorough cleaning and painting. A complete evaluation of the lift stations will be made by 7/1/94. This evaluation will serve as a tool in identifying needed repairs, inefficiencies, cleaning, and painting. I will have a complete report to the committee by the first time we meet in July.

Collection System

The collection system is being evaluated at this time. The information gathered by the smoke testing program will be our tool to identify I/I areas, and the backhoe attachment and supplies will be the means of repairing many of those areas. There will undoubtedly be areas we identify that will require a combining of resources from the Street Department, i.e., catch basin and storm line inflow. There will also be areas to repair that may be too deep for us to safely repair without proper shoring. These areas will be documented and a plan of action drawn up spelling out the best and most cost-effective ways to repair them.

EXHIBIT 3.2

City of Robinson
Five-Year Facilities Plan

Wastewater Treatment Plant

The design capacities of the plant (after installation of fine bubble diffusers) need to be analyzed for the Wastewater Treatment Committee to determine if future expansion is relevant. The determining factors shall be:

1. Current and past 12-month hydraulic and organic loadings to the facility.
2. The affect of I/I repairs on hydraulic loadings.
3. Future growth potential of Robinson.
4. The historical rate of growth of Robinson.

5. Monies available for plant expansion.
6. Grants available to help fund plant expansion.
7. Biomonitoring data collected in 1995.
8. TRE results if TRE is necessary.
9. Plans to meet existing sludge regulations and future amendments thereto.
10. Plans to meet more stringent effluent discharge standards if imposed by EPA.

If it is determined that the Wastewater Treatment Plant needs to be expanded and/or a new plant built, a plan shall then be developed and implemented to select an engineering firm to do a facilities study, analyze the study to determine the most cost-effective means of expanding or rebuilding, secure funding to finance the project, and then proceed with the construction project(s) procedures.

If it is the Wastewater Committee's decision to expand the life of the current facilities as they exist (after fine bubble diffusers), the plan shall shape up as follows. The plan will be broken down to address each strength or weakness of the existing facilities.

1. Contact Stabilization Reactors

With the understanding that when the new diffusers are installed in spring of 1995 that any needed painting or structural work is performed, those units should be set except for the following items. These items need to be taken care of in the first two (2) years of this plan.

1. The anodes, which keep the structures from deteriorating due to rust, should be inspected and replaced if needed. $2,000.00
2. The exterior of the tanks should be repainted after replacement of the anodes. $500.00
3. A spare clarifier scraper arm gear drive should be purchased in case of existing gear drive failure. $5,000.00
4. The Influent splitter box needs a new hinged lid enclosure fabricated and installed. $300.00
5. The RAS splitter boxes should be upgraded to be leakproof and to contain WAS flow measurement devices. $4,500.00

2. Influent Barscreen/Grit Collector

This existing equipment needs to be replaced with more efficient and effective Preliminary Treatment equipment. This will be very crucial because the ceramic fine bubble diffusers are prone to plugging caused from grit and debris accumulation. The new equipment can be sized and configured so it can be used for preliminary treatment in a new plant later in time.

1. Have an engineering firm perform a study on what types of equipment would best replace the existing.
2. The engineering firm would report back on the best and most cost-effective solution based on the following specifications.
   - Would have to remove both grit and objects greater than 1/2″ in size.
   - Would have to operate in automatic mode.
   - An access road would need to be included to enable a refuse truck to pick up the removed debris which would automatically be placed in a 4-yard dumpster.
   - If the Grit Chamber is included as a separate structure, it shall be erected as such that the grit would be removed by a mechanically operated clamshell device mounted on an I-Beam. Both Barscreen and Grit Chamber units shall be configured so they can be bypassed for a period of time for maintenance and repair activities. The bypass shall contain at a minimum, a manually raked barscreen.
   - The Barscreen shall be enclosed by a structure which is supplied with explosion-proof fixtures and which is large enough to allow maintenance and repair activities. Also, it shall be supplied with a heat source which will keep the screen and influent sampler from freezing in cold temperatures. $90,000.00

3. Wastewater Plant Influent Pump Station

The following items need to be replaced to expand the life of operations and safety of the lift station.

1. The #1 pump mechanical seal should be replaced and switched over to a grease lubrication system. $600.00
2. The manlift should be inspected by a professional to determine its safety and possible needed mechanical upgrades. $100.00
3. A spare sump pump should be purchased and "on hand" in case of emergency. $350.00
4. A spare mechanical seal should be purchased and "on hand" in case of emergency. $550.00
5. The pump impellers need to be inspected and replaced if necessary. If they do not need to be replaced, they may need to be reworked and balanced. $800.00

4. Rotating Biological Contactors (RBCs)

1. The RBC channels should both be drained and cleaned of any accumulated debris. $300.00 (Most can be performed by in-house labor.)

2. The electrical control box which houses the light switches needs to be replaced. $450.00
3. Any doors or windows which are defective need to be repaired and or replaced. (The Southeast door is missing.) $2,000.00

5. Tertiary Clarifiers

1. The fiberglass weirs need to be replaced. $4,500.00
2. The scum trough drain and water inlets need to be replaced. $1,000.00
3. The scrapers and drive mechanisms need to be inspected by a professional and determine if they need to be reworked or if they just need to be adjusted. $1,000.00
4. A spare gear reducer drive unit should be purchased and "on hand" in case of emergency. $500.00

6. Tertiary Sand Filters

1. The backwash pumps should be removed one at a time for rebuilding and/or replacement. $500.00
2. The proximity switches should all be replaced. $400.00
3. A chemical cleaning system should be installed to allow removal of grease and debris which normal backwashing does not remove. $5,500.00
4. The traveling bridges should be repainted. $400.00

7. Plant Effluent Flow Measuring/Sampling Station

1. A structure needs to be installed to house the effluent flow measuring transducer and effluent composite sampler. The structure needs to be heated and have explosion-proof fixtures. $5,000.00

8. Facility Blowers

1. All four blowers should be inspected to determine if tolerances are correct and the bearings are in good order. If repairs are warranted, they should be performed. $1,000.00
2. A fourth 75-horsepower blower should be purchased and installed. The fourth blower will provide backup air capacity and a backup blower if ever needed. (This would only be done if IEPA requires another blower to rerate the POTW design capacity.) $50,000.00
3. A permanent ventilation system needs to be installed which will utilize a large fan to circulate outside air through the blower room during the summer months. $500.00
4. The buried portion of air transfer line which sends the blower air to the C/S tanks needs to be replaced. The existing line has many leaks and

will more than likely become less efficient as time passes. The best way to do this is fabricate the new line from stainless steel, and route it overhead, above ground, to the C/S tanks. $35,000.00

9. Belt Filter Press

1. The existing press is operating well at this time, but needs to be inspected to determine if upgrades need to be made to increase its service life. This inspection report shall be our plan of refurbishing the press. $500.00
2. A new set of belts and doctor blades need to be purchased and "on hand" in case of emergency. $2,500.00

10. Sludge Processing and Disposal

1. Sludge production and disposal will be a major issue in the next 5 years. The reason is the newly imposed Federal 503 regulations which did not exist 2 years ago. The Federal regulations contain more sludge testing parameters and more stringent limits than did IEPA in the past. The parameters of concern, and their descriptions, are listed below.
   a. Vector Attraction Reduction
      This is a measure of the "volatile" or undecomposed organic material remaining in the sludge after digestion. It is expressed in terms of percent volatile solids. Since it has never been an IEPA regulated parameter, many Illinois POTWs are not able to meet the limits because of a lack of necessary equipment.

      If a WWTF cannot meet the limits by means of the first 8 options listed in the Federal Regulations, you are responsible to have the sludge incorporated into the soil within 6 hours of land application. (IEPA was 48 hours.) (If a farmer is responsible to incorporate the sludge, which he is in our situation, he may choose to not accept the sludge based on this criteria.)
   b. Pathogen Reduction
      This is a measure of disease-causing bacteria contained in the sludge. It is expressed in terms of Colony Forming Units (CFUs) per dry gram. Once again, this was an unregulated parameter by IEPA and was never before an issue. The Robinson POTW, like many other Illinois POTWs, are not currently equipped to remove the bacteria and are at the mercy of either passing the test or failing it. We have met the criteria so far, but are borderline. If the POTW sludge should not meet the criteria, it cannot be land applied. This means it is considered "Hazardous Waste," and if we

do apply the sludge, we will be in violation of the 503 Regs and face enforcement action.

Both of these parameters can be easily met by one simplified process: "Lime Stabilization." We would need to install equipment to mix powdered lime into our dewatered sludge off the Belt Filter Press. The lime must raise the pH of the sludge to 12 (expressed in standard units) for 2 hours and remain at 11 or above for 22 more hours. This process gives you what is referred to as Processes to Further Reduce Pathogens (PFRP). $180,000.00

I think we need to explore this option, as well as others in the immediate future as to ensure consistent compliance.

11. Lift Stations

1. All but three existing Lift Stations should be fine for the next 5 years. The three that are in need of upgrade are Lincoln Street, Pickford Street, and Tyler Street. These are the three oldest existing lift stations and are in need of new pumps and electrical controls. Below, I will break down the needs of each station separately. *Note:* If all 3 can be outfitted with the same model pump, we can purchase 1 extra to have "on hand" for emergencies.

Lincoln Street

- Needs 2 new drywell submersible pumps and controls.
- Needs a correct type building erected over it which will allow room for servicing.
- Based on current daily hour readings, the pumps may need to be larger.
- The inside of the station will need to be sandblasted and repainted.
- Needs new suction and discharge piping, check valves and isolation valves.
- The ventilation system needs to be upgraded. $25,000.00

Pickford Street

- Needs 2 new drywell submersible pumps and controls.
- Based on current hour readings, the pumps may need to be larger.
- A new access ladder needs to be fabricated and installed.
- The inside of the station will need to be sandblasted and repainted.
- Needs new suction and discharge piping, check valves and isolation valves.
- The ventilation system needs to be upgraded. $25,000.00

Tyler Street

- Needs 2 new drywell submersible pumps and controls.
- Based on current hour readings, the pumps may need to be larger.
- A new access ladder needs to be fabricated and installed.
- The inside of the station will need to be sandblasted and repainted.
- Needs new suction and discharge piping, check valves and isolation valves.
- The ventilation system needs to be upgraded. $25,000.00

Other Lift Station Needs

1. We should purchase an electric winch to be mounted on the 1-ton truck to use for pulling and re-installing pumps from drywells/wetwells. $2,500.00
2. We should purchase, and have "in stock" a spare mechanical seal for each type of pump in case of emergencies. $4,000.00
3. We should plan on replacing the 1-ton dump bed truck with a new one in 1996. The winch could be transferred over to the new truck. $20,000.00

12. Collection System

The collection system shall be broken down into separate sections.

Rolling Stock/Equipment

1. The water jet truck needs the rusted areas sandblasted and repainted. $500.00
2. We will need a new root cutting device for the water jet truck in 1997. $800.00
3. We should purchase our own televising system for use in I/I control, identification of sewer pipe failure, locating connections etc. $5,000.00
4. The 1/2-ton Chevy pickup should be replaced in 1997. $20,000.00
5. The Jeep pickup should be replaced in 1998. $19,000.00
6. A portable, yet properly sized shoring system must be purchased in order to repair excavated areas safely and properly. $7,000.00

Sewer Lines

1. Repairing identified I/I areas will be an ongoing project throughout the 5-year period. Listed below is each certain defect type and the associated manpower and resources which will be needed for repairs.
   a. Sewer mains

These sections can be repaired by digging up the pipe in the area which was identified by smoke testing and replacing and/or repairing the defective pipe. $29,000.00

b. Manholes
The manholes which were identified in the smoke testing as taking on surface water will need to be addressed on an individual basis. The manhole lids which have large pick holes need to have new seal-tite lids installed. Manholes which leak around the barrel sections need to be tuckpointed and if in bad enough shape will need a professional regrouting. Manholes with lids located below grade need to be raised with concrete extensions so they are at grade. $20,000.00

c. Stormwater inlet direct/cross connections
These will be the most costly of all. If separate storm sewers are available in close proximity, the storm water can be piped to those conduits. If there is no separate stormwater line available, new storm lines will need to be installed. This will require the use of an engineer and possibly a private contractor to install (depending on the scope of installation).

My rationale in development of this plan

Although this 5-year plan is very aggressive, we need to keep in mind that the next 5 years will be very crucial to stabilize the existing facilities and remain in compliance. By choosing to upgrade and/or expand the existing facilities, we are "sending a message" to the community that we have identified what our current strengths and weaknesses are. Furthermore, that we are willing to address and remedy each weakness by the most cost effective means.

I feel that if Robinson continues to grow at the rate it historically has, it is inevitable we will need to upgrade the plant capacity or build a new plant within the next 10 to 15 years. This rationale is based on "current" IEPA and USEPA limits and guidelines. If the guidelines become more restrictive (which is very possible), our plan may need to be modified to reflect the changes.

The majority of the 5-year plan upgrades will not be "money thrown away." If a new plant is constructed down the road, most of the existing facilities can be utilized for sludge processing. The individual lift station upgrades performed in the next 5 years should prove adequate in those specific sites for 15 to 20 years.

The current hydraulic loading capacity of the plant (gallons of water to the plant a day) may be increased by getting the plant "rerated." However, certain limiting factors will not allow the existing facility to exceed a certain point. If the current rate of residential and commercial development continues in Robinson, the remaining hydraulic capacity could be reached within 10 years. If a new industry and/or industries locate in Robinson which emit large volumes of wastewater, the remaining hydraulic and organic capacity could disappear even sooner.

If we implement this 5-year plan, we can stabilize the current facilities to comfortably meet our NPDES and Sludge disposal permit criteria. In the interim, I think we need to start planning for a major plant expansion.

We should be looking ahead for funding sources, preparing necessary documentation to secure the funding, have plans prepared to construct the new or upgraded facilities, and then proceed with construction.

The entire process of building a new plant is much easier for all involved when you have planned ahead, know what it will cost, know you can afford it, and know what the results will be when complete. If all of these steps are clear and concise, we should be successful in getting the community to "buy-into" our plan.

Thank you for allowing me to submit to you my 5-year plan.

If you should have any questions concerning this plan, or if you would like for me to be more specific on any items, please give me a call.

Sincerely,

CITY OF ROBINSON

Superintendent, WWTF

EXHIBIT 3.3

CITY OF ROBINSON WASTEWATER TREATMENT FACILITY
PROPOSED BUDGET FOR FISCAL YEAR 1996/1997

ITEM	LAST YEAR	PROPOSED 96/97
SALARIES	201,000.00	205,000.00
CLERICAL	15,000.00	15,000.00
UNIFORMS	1,500.00	1,500.00

ITEM	LAST YEAR	PROPOSED 96/97
GR. INSURANCE	8,813.00	12,000.00
D&A TESTING	0.00	1,300.00
CONTRACT SERVICES	40,000.00	35,000.00
LEGAL/AUDIT	20,000.00	10,000.00
PRINT & PUBLISH	300.00	1,000.00
POSTAGE & FREIGHT	7,600.00	6,600.00
INSURANCE & BONDING	30,000.00	30,000.00
UTILITIES	70,000.00	65,000.00
M&R	40,000.00	35,000.00

$ 10,000 for 50 seal-tite MH lids, rubber couplings, PVC pipe, resin impregnated mortar for MH tuckpoint & misc. supplies to reduce I/I. The other $ 25,000 for regular WWTF scheduled & unscheduled O&M.

ITEM	LAST YEAR	PROPOSED 96/97
OTHER	2,000.00	2,000.00
GAS & OIL	2,000.00	2,000.00
M&S	40,000.00	30,000.00
BILL PREPARATION	15,000.00	12,000.00
PURCHASE OF EQUIPMENT	165,000.00	19,000.00

Sewer TV unit, grease & sand nozzles, 1 FM 2-way radio, copier for maint., laser printer, NH3-N & pH probes, lift station cleaner/mixer and casting setter/puller.

ITEM	LAST YEAR	PROPOSED 96/97
CAPITAL IMPROVEMENTS	0.00	110,500.00

Retrofit new pumps and controls in Tyler St. LS, replace air transfer line from blower room to C/S tanks, 1 ton utility bed pickup truck to replace 88 Chevy pickup, & Sludge Storage building ($ 101,000.00)

ITEM	LAST YEAR	PROPOSED 96/97
TRAINING	3,000.00	4,000.00
BOND P&I	105,845.00	109,835.00
DEPRECIATION & AMORTIZATION	130,000.00	130,000.00
	----------	----------
Total:	$ 897,058.00	$ 836,735.00

CHAPTER 4

Soliciting Help

If you are skeptical or afraid to begin the upgrade process on your own, there is help available. There are firms specializing in repair or retrofit project management. You may have met some at seminars or conferences. Give them a call. They may have performed similar projects for others and could give a very close estimate of the costs and time frames for completion. Again, call neighboring operators to see if they have any ideas. I have found that fellow wastewater treatment operators are always willing to lend a hand if possible. "I also firmly believe that wastewater plant operators are some of the best people in the world."

Start out by hiring a firm to give an estimate or quote, formulate a schedule for the project, provide descriptions of each step, etc. It is very important to start a file for each project you address. Have the firm copy you on every document associated with the project, such as

- plans
- specifications
- bidding documents
- public notices for bids

These items are typically not copyrighted, and you can use them to format your next project if you choose to manage it yourself. It may be necessary to change project descriptions, dates and times, specifications, etc., but the format is there and will save time and money.

These specialized firms can also help in the bid-awarding process. They may have past experience with similar equipment and know which ones will last, are easier to maintain, contain better warranties, better service, better performance, etc. This is all very important because all equipment is not comparable. *Note:* Please ask for, and check references for any company you may want to utilize. The best check of credibility is word of mouth from previous clients.

SELECTING A CONSULTING ENGINEER

If you feel there is a need to select a consulting engineer for whatever phases or aspects of a retrofit project, there are a few rules of thumb to abide by. The ultimate goal should be to select the best firm that can deliver the project results on time and within budget. Some municipalities will Public Notice a Request for Statement Of Qualifications (SOQs). In this case, the city officials can review the SOQs and narrow down the list to maybe two or three firms. Typically, this is not necessary because it is not mandatory to bid out or make consulting firms compete against one another for the job. In fact, some say it is unethical.

A very important criteria to stress is, the consulting firm to be selected will need to be able to communicate, solicit background information, listen, be available, and network with key individuals of the municipal staff. When it comes to choosing a consulting firm, remember they are similar to equipment in that not all are created equal. The firm chosen should be matched to the size of the project. For instance, a project that is estimated to cost $20,000.00 may not require the resources of a large firm. Maybe a smaller, local firm will be fine. On the other hand, a project that entails a vast amount of restructuralization, process flow changes, construction or operating permits, variances in effluent limits, rerating the original design parameters of the plant, etc. will more than likely require a firm that has enormous resources.

A clear and concise description of the project scope should first be defined. In reality, for most typical retrofit jobs this is pretty simple (refer to sample specifications contained as Exhibit 4.1). Of course, the more complex the project is, the more detailed the specifications are required to be. Take, for instance, the Pump Station Retrofit Specifications in Exhibit 4.2. This is a relatively straightforward project that a single company is providing a turnkey result. The representative from whom the pumps and associated equipment were purchased is the same company that is installing the equipment. It is unnecessary to get into great detail on this project. On the other hand, for a medium to large project, a consulting engineer may be necessary to identify the project scope, develop concise specifications, provide on-site project management, oversee the project to ensure all aspects of the contract are being performed correctly, etc.

Other very important selection criteria include the following:

(1) The firm should specifically identify who will be responsible for executing the different phases of the project (Senior Project Engineer, Senior Engineer, Resident Engineer, etc.).
(2) The city should be informed as to how they can be contacted, their availability to be involved, and the time they will each devote to the project.
(3) The individual(s) from the firm that will directly communicate with the municipality's key individuals should be

knowledgeable in the field, easy to work with, capable of listening, exchange ideas, be diplomatic, etc.

"There is no room for swelled heads here." If all of these steps are followed, an appropriate engineering firm will typically be found.

ORIGINAL EQUIPMENT MANUFACTURERS

If you have equipment that is not treating over design capacity but is failing due to wear and tear, you may be able to replace it with an identical item. Equipment manufacturers tend to change designs over periods of time but rarely completely redesign equipment so it would not retrofit into the same structure. Call the equipment manufacturer and find out if this is the case. If it is, money and time can be saved a direct retrofit of equipment.

Because the plant design will not be changed, it will probably not be necessary to obtain a construction permit from your regulatory agency. This deletes the necessity of associated engineering costs.

If by chance the original manufacturer has gone out of business or was bought out by another company (which is very possible depending on equipment age), it may be necessary to use similar equipment manufactured by another company. They may be able to fabricate their equipment to mount into the existing structure. Also, because the project requires an identical item (which can only be acquired from a single source), it may not be necessary to bid the project out. I am a firm believer that one should try to keep equipment designs as simple as practical. Typically, the fewer high tech accessories (bells and whistles) contained, the easier and cheaper the equipment will be to operate and maintain. I have seen many plants that contain high tech telemetry, computerized timers, processes operated solely on readings from fixed dissolved oxygen probes, pH meters, etc., and these are fine as long as they are necessary and actually work. However, I have seen these same types of equipment in plants where they are not necessary, and they are driving operators up the wall. Typically, the normal facility electrician cannot work on them due to their complex solid-state circuitry. If you call in a factory technician to repair the equipment, it may take days before they can get you in their schedule, and when they do, look out when the invoice arrives. I do know that process analyzation and electronic equipment has come a long way in the past few years, and there are certain manufacturers better than others. If it is necessary to install the high tech equipment, get references from the manufacturers and then contact the references to find out what their experiences have been. It is also a good idea to verify and specify that the equipment will work in a manually operated mode if necessary in case conditions should require it to do so.

EXHIBIT 4.1

Specifications
Dry Well Submersible Pump Retrofit
Tyler Street Lift Station
Robinson, Illinois

Part 1.0—General

Scope

Provide all necessary materials and equipment to retrofit the existing duplex drywell/wetwell Tyler Street Lift Station to a duplex dry-pit submersible pump station, new valves, piping, control panel and float switches.

Submittals

Submittals shall include at a minimum an equipment list and information containing; pump manufacturer, pump O&M manual, pump curves, pump technical specifications and warranty information, valve manufacturer, valve manual, valve specifications, control panel manufacturer, control panel manual and control panel specifications.

Products

Two (2)—dry-pit submersible pumps:

The pumps used shall be FLYGT CT 3127, or equivalent.
(7.4 HP, 230 Volt, 3 Phase, 4″ × 4″, 325 GPM @ 15′ Tot Head)

Four (4)—4″ plug valves shall be Pratt, or equivalent.

Two (2)—4″ swing arm check valves shall be Pratt series 18, or equivalent.

One (1)—Electrical control panel shall be, or equivalent to; Consolidated Electric, Model F100 with NEMA Type 3T (Weatherproof/Tamperproof) enclosure made of galvanized steel, painted, and suited for wall mounting. Controls must contain at a minimum: 4 float operation, pump alternator, hour meters, HOA switches, pump run lights, pump seal fail indicator lights, pump thermal overload indicator lights, phase loss safety cut off and lightning protection. Also, two (2) sets wiring diagrams & schematics.

One (1)—Stainless steel wetwell float switch holder.

Four (4)—Wetwell float switches.

All necessary associated anchors and hardware: Shall be HILTI, stainless steel, or equivalent.

All necessary piping and fittings: Shall be schedule 80 PVC.

Spare Parts

Two (2)—Mechanical seals (one upper, one lower).

Warranty

Warrant all parts and materials for a period of five (5) years after installation and start-up.

To repair defects in parts and materials found to be defective within the warranty period.

Manufacturers

Acceptable alternate dry-pit submersible pump manufacturers include Hydromatic and Yeomans.

Acceptable alternate manufacturers of associated plug valves, check valves and electrical controls/panel shall be at the City's discretion. (Please specify manufacturer of these items.)

EXHIBIT 4.2

Specifications
Dry Well Submersible Pump Retrofit
Tyler Street Lift Station
Robinson, Illinois

Part 1.0—General

Scope

Removal of all existing valves, pumps and pump controls, with disposal by the City; furnish all labor and equipment to install two dry pit submersible pumps, 4″ plug valves on inlet lines of drywell, 4″ discharge plug valves on discharge of pumps, 4″ swing arm discharge check valves on discharge side of pumps, all associated piping, float switches and electrical control panel in the lift station.

Replacement of the pumping equipment shall be performed in a sequence so as to minimize interruption of normal operation of the station. Also, to maintain bypass pumping operations as needed.

Safety

All work shall be performed in accordance with OSHA standards.

Notes

The existing equipment shall be removed and placed on the outside of the station for City employees to dispose of.

Any bypassing of the lift station shall be pumped to the upstream manhole located approximately 650′ from the station. The installing contractor shall be responsible to provide an adequate size pump, float system and necessary temporary piping to facilitate bypass operations.

All contractors are required to sign an Indemnification Agreement holding the City harmless during its performance of the contract if awarded to them.

Warranty

Warrant installation workmanship for a period of one (1) year after installation and start-up.

To repair defects in workmanship found to be defective within the warranty period.

CHAPTER 5

Obtaining Necessary Funding

Because the availability of money is usually the biggest obstacle in project initiation, considerable thought should be focused in this area. If it is not possible to fund the project(s) via the fiscal budget, you may need to solicit help. The first option to explore is federal or state assistance. Call your state environmental agency and ask for literature and information for funding assistance. There may be matching grants (50/50), low-interest loans (State Revolving Loan Funds), assistance for economic development projects, etc. You may also choose to issue tax-exempt municipal bonds. There is no limit to the sources you can tap into. I recently signed on the internet (World Wide Web) and have found web sites from state and federal governments that list these options. Here are a few different sites that have information concerning funding and/or grants.

- http://es.inel.gov (USEPA Enviroene)
- www.epa.gov/OWM (USEPA, Office of Wastewater Management)
- www.ilcommerce.com (Illinois Department of Commerce and Community Affairs)

There are numerous other WEB sites that are designed to help operators, managers, superintendents, engineers, etc. They contain discussion groups in which an individual can post a question concerning process control, collection system, sludge management, etc. Professionals from around the globe respond to the questions by posting relevant answers. They also contain listings of other wastewater treatment plants across the globe, their associated e-mail addresses, facility information, contact person name, phone numbers, and so on. Below is a list of a few excellent WEB sites.

- www.wef.org (Water Environment Federation)
- www.www.com (Water and Wastewater Web)
- www.caeconsultants.com (CAE Consultants, O&M News)
- www.wateronline.com (Water Online)

Of course, these are only a few of the numerous resources available on the Internet and World Wide Web. Another valuable resource that will soon be available in our state is Illinois Environmental Protection Agency's Title 35, Environmental Protection Regulations. If a section of the regulations is needed, visit their site and download the required information.

GOVERNMENTAL FUNDING (REVOLVING LOAN FUNDS)

These types of monies usually require significant amounts of paperwork to qualify for. You may also have to compete for funds that are available. However, if the need is great enough and it is the only realistic way to finance the project, then it is well worth the extra time and effort to apply. Typically, the necessary paperwork will consist of a facilities plan (an engineering firm would need to put this together), accounting data from the specific municipality and department, and a committed council or board to achieve this goal. There may be someone in your community who is very good at applying for grants and assistance (local chamber of commerce, politicians, economic development groups, etc.). Don't be afraid to contact these types of individuals and ask for help. Regardless, the typical interest on the loan can range from 0 to 3%. These are very good interest rates and will keep your payback stress at a minimum. The federal government appropriates these state revolving funds each year. It is constantly a battle for the politicians to agree on what, or if any, state revolving funds will be appropriated and budgeted for the fiscal year. As long as government remains committed to upgrading and updating the country's infrastructure to protect the nation's waters from environmental degradation, this money will be available.

A study of municipal income in terms of user revenues will also be required. The lending authority must have this information to ensure your rates are adequately structured to generate sufficient money to retire the loan. If the rates are determined inadequate, the municipality must then amend their ordinance(s) to raise rates. This is another issue that will get immediate attention from the public. This is why everything in the planning process must be justified as well as being relayed to the community. People generally do not have a problem paying a little more if they feel they are receiving a benefit in return.

ISSUANCE OF BONDS

Tax-exempt municipal bonds are probably the simplest way to fund municipal projects. This is normally the case unless the municipality is currently overburdened to retire existing bonds for past projects. If the need has arisen to use these sorts of bonds as financial aid, it would be a good idea to contact the municipality's law director (city attorney) and the city treasurer or financial director. Find out whether you are in the process of retiring existing bonds and, if so, what the payoff amount is. If you can pay off the existing debt (with cash reserves, savings, etc.), it is best to do so to avoid borrowing beyond your means. In most cases you must also publish the city's intent to issue bonds in the newspaper. The criteria differ, but typically there is a month-long waiting period in which the public can comment on the necessity of the bonds.

If there is no resistance to the bond issue, the city can move forward with the selection process of the bond type to pursue. On the other hand, if there is public comment and scrutiny, the issue will need to be included on a ballot in the next general election. In this case, the election results will identify the city's ability to proceed with or stop the bond process. This is a good example where public image, support, and communication of the municipalities staff, as well as public officials, is vital. Let's say, for instance, your department had a track record of excessive spending while providing less than desirable results. I guarantee that if this is the case, the referendum issue will perish when voted on. However, let's say you have proven to operate a cost-effective facility, the department has a very good image in the public's eye, as well as portraying a proactive image. The requested project has been communicated thoroughly to the public and the project seems rational. More than likely, the bond issue will pass. Other options may include contacting local banks and lending institutions to see if they would be interested in issuing bonds. Bonds are a very good investment, and local banks or firms may be more than willing to accommodate your needs.

Because it is unnecessary to jump through as many hoops for bonds, as opposed to revolving loan funds, they are easier to obtain in terms of paperwork, comprehensive studies, and so on. However, they will usually contain a higher interest rate than the SRF monies.

CHAPTER 6

Permits to Construct or Operate from EPA

Most state regulatory agencies require a permit application and permit to be granted before an upgrade can proceed. This may not be the case for a simple equipment retrofit. If you are not changing process design or flow patterns, a permit will probably not be necessary. Call your state regulatory agency's permit section and ask. It is a good idea to get it in writing if possible. They may need a project description, preliminary drawings, etc., but that is still less expensive than a full-blown permit application put together by an engineering firm. The general rule of thumb is: Any permit application for operating (NPDES, Biosolids Application, etc.) does not require an engineer's seal for submittal to EPA. Also, any project that does not change the capacity, flow schemes, or flow patterns does not require a permit. The operating permit applications do require quite a bit of facility data to meet requirements. The data may be found from a previously submitted NPDES or Land Application of Sludge permit application from a previous engineering firm. If there are no copies of these previously submitted applications, request them from the engineering firm.

Another rule of thumb is: For any type of new construction

(e.g., installing another clarifier, new pumping station, etc.), a permit will probably be necessary. Retrofitting equipment into an existing structure (e.g., barscreen, clarifier scrapers, pumps, etc.) may not require a permit.

Any necessary bidding documents for installation by a contractor (drawings, specifications, etc.) may be provided by the equipment manufacturers. Typically, they will be happy to provide this information if they believe they have a good shot at selling you their equipment.

If a permit is required, there are a few helpful hints to save money:

(1) Find out which equipment manufacturers wanting to bid on the project are willing to provide you with all necessary drawings, specifications, basis of design data, etc. They usually have engineers on staff who can put these documents together for you free of charge (by the way, this is usually where engineers get their drawings, specifications, etc.).
(2) If the manufacturer is from your state, its engineering seal should be satisfactory for your application.
(3) If not, ask an engineering firm from your state to review the basis of design documents and drawings for a fee.
(4) If the equipment manufacturers' data all meet the design criteria by the state, the engineer should sign and seal the documents.

These are all procedures and steps that may be taken care of by equipment rehabilitation professionals if you do not feel comfortable doing it yourself.

Once all permits and necessary contract documents are in place, you can proceed with public notice for bids. Generally, most publicly owned treatment works are required to solicit bids for projects that will cost over $10,000.00. These criteria may vary according to state or local law. If you are not sure, contact your municipal law director (city attorney). Sometimes in emergency situations, you may not need to bid a project out. Usually it is best to do so to avoid the connotation of playing favorites. Also, a competitive bid creates an environment in which all bidders "sharpen their pencils" to submit their most competitive bid. Even if you know who will probably get the bid, don't let the bidder know, to keep the bidding competitive.

CHAPTER 7

Organizing Teams

Because teams and teamwork are mentioned quite a bit in this book, I felt it would be a good idea to expand more on this issue. A definition of *teamwork* could be a collaborative, unselfish effort by a group of individuals, which as a result, accomplishes a common goal. This definition by no means counts out individual efforts of persons. Individual accomplishments are teamwork oriented as long as they are directed toward the common goal as well. However, individual efforts that are for the sole purpose of individual notoriety can be demoralizing and destroy the working environment of a team. Let's say, for instance, that a facility superintendent has stressed "teamwork, teamwork, teamwork" and subsequently accepts all the credit for the results that are produced. It doesn't take long for his concept to flop. He will lose all respect from his subordinates if he continues to behave in this manner. Teamwork as a rule must not only be preached but practiced as well.

A good outlook for supervisors to have is: "Things are to be managed; people are to be lead." A supervisor must not only be a manager of things (administrative tasks, budgets, maintenance programs, safety programs, etc.) but should also be a leader of people. Like a coach on a football team, he is an extension of the staff. He sets examples, he shows restraint, he is compassionate, empathetic, rational, etc.

Another deleterious effect on teams is not treating all employees fairly. This type of activity can destroy employee morale faster than any other. Let's say the city has strict tardy policy, and the employees all know very well what the policy is. An employee who does not have the best work ethic has broken the rule three times in a month. Disciplinary action is taken, and he is suspended for 3 days without pay. Six months later, your highest output employee comes in to work late three separate times in a month. Nothing is said to him because of his great work ethic. A message has just been sent to all employees that the supervisor is not fair. He has once again lost respect from his subordinates. The best mottos to use are: "Be firm, but fair," and "Practice what you preach."

Supervisors need to set good examples for their employees as well. If a supervisor requires everyone to be at work at a certain time, then he needs to be there at that time, and better yet, earlier.

It is a good idea to have facility meetings whenever possible on a regular basis, such as maybe the first Monday of the month or even maybe once a week. This is a good time for supervisors to update their team(s) of recent happenings at the managerial level. It is also a good time for employees to air differences, ask questions, provide input in terms of ideas, speak about accomplishments, etc. The more of these meetings that are held, the more comfortable the employees will feel to opening up. Don't be discouraged if the first few meetings do not yield many results; as time goes on, the dialogue will begin to flow.

Anyone who has a facility that runs on a legitimate "teamwork" basis will tell you it is the best thing since sliced bread. Sure, you will have little problems here and there, because that's human nature. However, the ability of a crew all working toward the same goals is phenomenal. For instance, you know how it feels when you and your friends are engaged in a task everyone enjoys (boating, fishing, hunting, camping, etc.). It's fun, and time really flies. That is the same feeling a team will get when they are working on a project together. Even if it is a tough job, nobody really cares because everybody is pitching in and working toward the common goal.

Recognition is another big part of teamwork. When a project is completed, the team as well as superior individual efforts should be recognized. A good old "pat on the back" can be better for morale than slipping a $100.00 bill in their pocket. It will make them feel good about themselves and the team. On the other hand, if an employee is slacking off and not a team player, this must be addressed as well. The employee must be taken off to the side and informed of inadequacies to maintain equilibrium. It is a good idea to tell employees up front that they should not take actions like these personally, but, instead, realize it is necessary to maintain a stable work environment. Good teams always share ideas and express feelings and concerns among the group. There is no room for hidden agendas or harbored feelings. The feeling of uninhibited expression of one's feelings is crucial to the team concept.

Another important part of teamwork is training. A balanced training mentality should consist of both on-the-job, and academic training courses, seminars, conferences, etc. Again, being fair in how the training is equaled out among employees is a juggling act. Because many smaller municipal operations have only a few employees, they cannot afford to send all of their people to every training event. The best way to handle this scenario is first to cross-train as many, if not all operators and maintenance people so they can spell each other in the other's absence. Try to get all employees involved in seminars and activities. If two employees are sent to a pump seminar, then send two other individuals to an activated sludge seminar. Upon returning, have a meeting with all employees and allow the attendees to share what they learned with the other staff members.

Superintendents, managers, lead operators, and public officials should all make it their highest priority to train and continually retrain their staff. They should also make it high priority to get the staff, as well as themselves, certified to the highest possible degree. A good way to promote these types of individual challenges for employees is to offer incentive packages that reward each level of certification achieved. The incentive may be in the form of cash payments, extra vacation, gift certificates, or a night out with the spouse at a nice restaurant, etc. Not only do these types of incentives entice them to achieve certification but also they promote a healthy competitive spirit among the team. Every situation is different, and there is no one specific way to prescribe teamwork. However, there are certain conditions, such as the ones listed above, that must exist for the teamwork concept to work.

CHAPTER 8

Public Notice for Bids Request

Most bid notices are placed in the Public Notice section of the largest newspaper circulating in the community. Again, this is subject to state and local laws. When the ad is printed in the newspaper, a lengthy write-up, including specifications, instructions, etc., is not necessary. Those items can be referred to as a package prospective bidders may pick up from the project manager. It is also customary to charge a fee for these items to recoup any administrative costs associated with putting them together, e.g., typing, copying, postage, etc. Depending on the magnitude of the project, the cost could be anywhere from $20.00 to $100.00.

If there are certain manufacturers and/or contractors you want to bid on the project, send them a bid package directly. I usually do this because we have no law specifying how many times the ad is to be placed in the newspaper. Depending on the size of the project, typically once is sufficient. If those you want to bid on the project miss that one-time publishing, they will not know to bid on it. It is also a good idea to make the specifications "very tight" to discourage inadequate or undesired bidders. This can be done by inserting state-of-the-art design requirements, extended warranty periods, etc. in the specifications.

Let's face it, it is usually not a requirement to award the project to the lowest bidder unless a funding source requires it, so make sure you are not stuck with equipment that has not proven itself or has shown to be susceptible to premature failure in similar installations. Be sure to include a statement in the "Notice for Bidders," which reads "The (insert your entity name) reserves its rights to accept the bid that it believes to be in its best interest, waive any formalities, and to reject any and all bids" (see sample contained as Exhibits 8.1 and 8.2).

It is also necessary to specify provisions for the bid to include two or three (however many you would like) copies of the following:

- equipment drawings
- detailed operation and maintenance manuals
- electrical drawings and schematics
- recommended spare parts lists
- spare parts
- maintenance schedules
- recommended lubricants

If these items are not specified in the bid package, it will probably cost an additional fee to get them at a later date.

AWARDING THE CONTRACT

When the date for bid opening has arrived, be present for the event. There may be bidders present for the opening as well. It is best to not make any comments concerning the bids during the bid opening while bidders are present. Save any comments until the bids and associated documents have been thoroughly reviewed.

This will allow time to make sure all bidding requirements and provisions were met. Once again, you may elect to hire an equipment rehabilitation professional to review the bids with you to determine if all criteria have been met. They should also have a good idea of which equipment has historically performed good or bad. Once the bids have been reviewed, the winner may be selected. Keep in mind that you have reserved the right to reject any and all bids.

If you feel good about a certain bid and want to award them with the project, then convey this information to your governing board. Let them know whom you would choose and why. If certain bidder(s) did not meet warranty provisions, equipment specifications, etc., the board needs to know this information in case they are asked. Even if the chosen bid was not the lowest, it still may be best due to certain conditions. Remember that this equipment will be around for potentially 20 years or more, so it needs to last.

FORMAL CONTRACT OF AGREEMENT

Once the bid has been awarded, a standard contract of agreement between the municipality (owner), and the equipment manufacturer or representative should be filled out and signed by both entities. The contract should detail what everyone's role is. (The standard form of agreement we use can be obtained from The American Institute of Architects, Owner Contractor Agreement, Construction Management Edition, 1735 New York Avenue, NW, Washington, DC 20006.) Certain specifics, such as

- equipment descriptions
- equipment delivery dates
- warranties
- start-up assistance

- operation and maintenance manuals (usually want two or three sets)
- drawings
- electrical schematics (two or three sets), etc.

should be spelled out in the contract. You may want to formulate your own agreement; however, if you do, let your municipality's law director check it out to make sure it contains all necessary verbiage.

If the equipment is to be installed by someone other than facility staff or the equipment manufacturer and installation is bid out separately, then another agreement needs to be filled out and signed by the municipality and the installing contractor. Certain specifics should be included in these specifications and contracts as well. It is a good idea to address warranty on quality of installation (typically the same length of warranty of the equipment), roles of all parties during installation, such as

- tank cleaning responsibilities
- site cleanup
- site landscaping subsequent to project completion
- start dates
- completion dates
- penalties for late completion
- coordination of unit processes being off- or on-line
- painting of equipment, etc.

If these things are important or required by the municipality, they should be specified. It is also a good idea to check if there are laws governing, liability insurance coverage requirements for contractors, prevailing wage requirements, hold harmless agreements, bid bonding, performance bonding, safety provisions, etc. These can make a big difference on someone's final bid price, and if they are not specified, the contractor is not subject to perform them.

PROJECT OVERSIGHT AND MANAGEMENT

This is another important task that requires time and resources. Dependent on the scope of the project, assistance may not be necessary. However, if you do not feel comfortable with this role, hire an equipment specialist or local engineering firm to perform this task. Have them submit a quote spelling out what it will cost to perform this task. If the project is very technical in nature or will be too time-consuming, better results may be obtained by passing the responsibility on to other capable individuals. Once again, if the task seems doable, then money and time can be saved by performing them in-house.

EXHIBIT 8.1

Public Notice

The City of [insert name] is now accepting bids for the installation of new pumps, controls, valves, etc. (all equipment provided by others) in the Tyler Street Lift Station. Tasks include: removal of all existing valves, pumps and pump controls, with disposal by the City; furnish all labor and equipment to install two dry pit submersible pumps, 4″ suction plug valves, 4″ discharge plug valves, 4″ swing arm discharge check valves, all associated piping, float switches and electrical control panel in the existing lift station.

The equipment will be provided by others.

Proposals will be accepted until [time, day, month, year] at the [city hall, street address, city, state, zip code]. Sealed bids must be marked "Bid for Lift Station Installation."

Specifications are available at the [location] [phone no.] or at the [location] [phone no.].

The City of [insert name] reserves its rights to accept that bid which it believes to be in its best interest, waive any informalities, and to reject any and all bids.

Specifications
Dry Well Submersible Pump Retrofit
Tyler Street Lift Station
Robinson, Illinois

Part 1.0—General

Scope

Removal of all existing valves, pumps and pump controls, with disposal by the City; furnish all labor and equipment to install two dry pit submersible pumps, 4″ plug valves on inlet lines of drywell, 4″ discharge plug valves on discharge of pumps, 4″ swing arm discharge check valves on discharge side of pumps, all associated piping, float switches and electrical control panel in the lift station.

Replacement of the pumping equipment shall be performed in a sequence so as to minimize interruption of normal operation of the station. Also, to maintain by-pass pumping operations as needed.

Safety

All work shall be performed in accordance with OSHA standards.

Notes

The existing equipment shall be removed and placed on the outside of the station for City employees to dispose of.

Any bypassing of the lift station shall be pumped to the upstream manhole located approximately 650′ from the station. The installing contractor shall be responsible to provide an adequate size pump, float system and necessary temporary piping to facilitate bypass operations.

All contractors are required to sign an Indemnification Agreement holding the City harmless during its performance of the contract if awarded to them.

Warranty

Warrant installation workmanship for a period of one (1) year after installation and start-up.

To repair defects in workmanship found to be defective within the warranty period.

EXHIBIT 8.2

Public Notice

The City of [insert name] is now accepting bids for all equipment and materials necessary for the upgrade of its Tyler Street Lift Station to a dry-pit type submersible pump station.

Proposals will be accepted until [time, day, month, year] at the [city hall, street address, city, state, zip code]. Sealed bids must be marked "Bid for Life Station Equipment."

Specifications are available at the [location] [phone no.] or at the [location] [phone no.].

The City of [insert name] reserves its rights to accept that bid which it believes to be in its best interest, waive any informalities, and to reject any and all bids.

Specifications
Dry Well Submersible Pump Retrofit
Tyler Street Lift Station
Robinson, Illinois

Part 1.0—General

Scope

Provide all necessary materials and equipment to retrofit the existing duplex drywell/wetwell Tyler Street Lift Station to a duplex dry-pit submersible pump station, new valves, piping, control panel and float switches.

Submittals

Submittals shall include at a minimum an equipment list and information containing pump manufacturer, pump O&M manual, pump curves, pump technical specifications and warranty information, valve manufacturer, valve manual, valve specifications, control panel manufacturer, control panel manual and control panel specifications.

Products

Two (2)—dry-pit submersible pumps:

The pumps used shall be FLYGT CT 3127, or equivalent.
(7.4 HP, 230 Volt, 3 Phase, 4″ × 4″, 325 GPM @ 15′ Tot Head)

Four (4)—4″ plug valves shall be Pratt, or equivalent.

Two (2)—4″ swing arm check valves shall be Pratt series 18, or equivalent.

One (1)—Electrical control panel shall be, or equivalent to; Consolidated Electric, Model F100 with NEMA Type 3T (Weatherproof/Tamperproof) enclosure made of galvanized steel, painted, and suited for wall mounting. Controls must contain at a minimum: 4 float operation, pump alternator, hour meters, HOA switches, pump run lights, pump seal fail indicator lights, pump thermal overload indicator lights, phase loss safety cut off and lightning protection. Also, two (2) sets wiring diagrams and schematics.)

One (1)—Stainless steel wetwell float switch holder.

Four (4)—Wetwell float switches.

All necessary associated anchors and hardware: Shall be HILTI, stainless steel, or equivalent.

All necessary piping and fittings: Shall be schedule 80 PVC.

Spare Parts

Two (2)—Mechanical seals (one upper, one lower).

Warranty

Warrant all parts and materials for a period of five (5) years after installation and start-up.

To repair defects in parts and materials found to be defective within the warranty period.

Manufacturers

Acceptable alternate dry-pit submersible pump manufacturers include Hydromatic and Yeomans.

Acceptable alternate manufacturers of associated plug valves, check valves and electrical controls/panel shall be at the City's discretion. (Please specify manufacturer of these items.)

CHAPTER 9

Communicating Results

After the project is completed and you are enjoying the benefits of upgraded or retrofitted equipment, you must communicate the positive results to all individuals and groups involved (see sample contained as Exhibit 9.1). The governing body is especially important due to their role in the decision-making process. These individuals have gone with your plan, and you are building a trusting relationship with them. You should invite them to your facility and show them firsthand the results you have achieved. Listed below are items that may have been affected.

- returned to compliance with your NPDES permit
- reduced or eliminated a labor-intensive job
- savings on energy (electricity, gas, water, etc.)
- reduced chemical use (polymer, chlorine, lime, etc.)
- reduced or eliminated odor complaints
- reduced overtime costs
- reduced administrative costs
- eliminated costly downtime

You may not realize it, but these results are very important and do make a difference. For instance, if you install a new fine

RETROFIT FLOWCHART

13. COMMUNICATE RESULTS TO THE PUBLIC

12. RETROFIT OPERATIONAL

11. PROJECT OVERSIGHT AND MANAGEMENT

10. AWARD BID, EXECUTE CONTRACT(S)

9. REVIEW BIDS

8. PUBLIC NOTICE REQUEST FOR BIDS

7. FORMULATE BIDDING DOCUMENTS

6. DEVELOP SPECIFICATIONS

5. OBTAIN NECESSARY PERMITS

4. OBTAIN FUNDING (IF NECESSARY)

3. SOLICIT HELP (IF NECESSARY)

2. PLANNING

1. EVALUATION OF EXISTING EQUIPMENT

pore activated sludge diffuser system and it allows you to maintain acceptable dissolved oxygen readings with one less blower on-line, this may save hundreds to thousands of dollars a month. You should translate these types of results into dollars and relay the information to all involved.

The best way to convey this information to your board is via a written monthly, quarterly, and/or yearly reports (see samples contained as Exhibit 9.2 and Exhibit 9.3). However silly this may sound, it works! I have found it best to report on a monthly and yearly basis. The reports can contain items, such as

- facility safety
- operations
- biosolids application
- maintenance and repair expenses for the month (you can itemize or maybe just give a total)
- descriptions of equipment repairs
- graphics, such as actual effluent quality versus permit limits, etc.

Do not just report on the good; you must report everything. The problem with only reporting positive aspects of your operation is that the board seldom recognizes you are having problems. They will feel much more comfortable with your opinions and suggestions if they are getting the whole story. They will realize you are capable of identifying and rectifying problems and issues. Most importantly, include information concerning upgrading or retrofitting results.

Don't forget to thank them for providing necessary funding for a project that has saved you time, money, and headaches. If they know you appreciate the help, they will be much more likely to continue to fund your projects. Some may not care and not read the reports, but others will and take pride in knowing they were an integral part of the final result.

Community support is a very important item as well. When you have completed a project that has yielded great results, have an open house and invite the public to see firsthand what you've done. After all, these individuals are the ones footing the bill. Advertise the open house in the local newspaper and give a brief description of why you are having the open house. Indicate the address of your facility (you would be surprised how many people do not know where your wastewater treatment plant is located) and last, but not least, get your entire crew involved. Let them help in conducting tours. If the project(s) were performed with in-house labor, they will be proud to show off the final results. This is a great morale booster. Invite local media to be present. We all know they chomp at the bit to report negative items; challenge them to report on positives. You will not change everyone's perception of your facility's integrity, but community support can start slowly and blossom over time.

A very important thing to remember is municipal governments (not all, but most) in the past developed a stigma of "underworked and overpaid." If this is the case in your community, you must break this stereotype and prove you want to operate your facility efficiently and cost-effectively. If you accomplish this, you have won the battle. Believe me, if you do not take the initiative of running an efficient operation, there are hundreds of contract operations firms who are. I would estimate that 90% of contracts awarded to private firms are due to one or more of the following reasons:

- POTW regulatory compliance problems
- process equipment failure due to inadequate maintenance
- poor budget performance
- poor planning
- personnel problems
- distrust of wastewater treatment staff

If a contract operations firm contracts your facility, you can rest assured they will address these issues. They are in business to operate wastewater plants as a business. They will capitalize on any chance to save energy (electric, gas, potable water, etc.), manpower, chemicals, etc. This is how they can operate the same facility as yourself, for about the same yearly budget and still make a profit.

The reason I know this is because I've worked for both private companies and municipal governments. Some of their best clients are those facilities which choose not to remain proactive and operate as a business.

EXHIBIT 9.1

"Press Release"
Sand Filter Renovation
Total Costs by City vs Bid

The minimum work bid recommended would have cost: (Parts only)	$58,014.00
City did the work in-house for: (Parts only)	$15,366.17
Savings:	42,647.83

The Sand Filter Renovation was a major success for the City. Ineffective filtering operations in the past had left the Wastewater Plant's effluent vulnerable to violations of the NPDES Permit Total Suspended Solids (TSS) Limit. The three major problems with the existing system were;

1. The filter media contained excessive amounts of grease and organic material which would not allow the water to drain through quickly enough.
2. The water sent to the filters contained too many suspended solids which clogged the already slow draining media.
3. The indexing backwash mode was not operational and did not allow for effective backwashing operations.

The City renovated the Sand Filters by utilizing the following resources and techniques.

a. The existing media was removed by "vacuuming" it from the filter cells. This operation was performed with the aid of a vacuum truck and operator rented from MAECO. The hardest part of the job was performed by City personnel. This operation spearheaded and led by a WWTF employee consisted of manually directing the vacuum hose into each cell and ultimately removing 100 tons of sand and gravel.

b. Once the old media was removed, the next step was to "super-chlorinate" and clean the underdrain system. This process was also performed by City personnel.

c. Once the Underdrain system was clean and under free flow conditions, it was time to install the new media. This process brought about many innovative and creative ideas. First, a plan to fabricate a chute which would ride on the traveling bridges and direct the media into the cells was developed and constructed by Superior Welding. This saved a tremendous amount of back-breaking work and time.

   The process of installing the new media was spearheaded by WWTF employees. They devised a system in which the most media could be installed per day, and manual lifting by facility personnel was minimized.

   The media installation took every WWTF employee to perform. Just to put this task in perspective, 6 men handled 2,000, 100 pound bags of sand/gravel (100 tons). The media was brought to the Filter building with a rented fork truck, the pallets were then pulled by personnel on a pallet jack (which was provided to us) to the area to be installed. The bags were lifted by 2 men up on a 2″ × 10″ board and slid to the man who was opening the bags and sending the media into the chute. A man was in the filter bed who was directing the media through a 4″ hose into the cells. Once the media was in the cell, he was required to level that layer before the next layer could be installed.

d. Once the new media was in, we concentrated on the electrical, mechanical, and physical integrity of the filters. We asked [insert name] to fabricate new aluminum support structures which hold the backwash hoods. They fabricated the structures and installed them.

Once the structures were sound, we began analyzing each electro-mechanical piece of equipment. Work orders which contained manufacturer recommended tasks and lubricating techniques were generated on all associated equipment. Once all bearings were greased, chains lubricated, gear oils changed, and backwash hoods aligned, we decided to put the filters "on-line" so we could check the remaining operations.

If was determined that certain electrical controls were defective after a trial run was made. New proximity switches, and control relays were ordered and installed by WWTF personnel to ensure the filters would backwash in "indexing" mode. Quite a few minor adjustments were made to ensure proper operation of the traveling bridges. Once this was complete, the filters operated as designed.

I would like to give credit where credit is due.

First, I would like to thank the Mayor and City Council for having the confidence in us to take on this task. Especially the Wastewater Treatment Committee members. If not for their timely approval, we would still be sitting here trying to decide how to pay for the expensive alternative. Also, we would not have the filters available for operation in case of imminent Suspended Solids Permit Violations and Biomonitoring which starts in July 1995.

The WWTF employees, especially the individual effort of John Doe who made it his personal mission to "get the job done." Other employees [insert names] all did a fantastic job. What we need to remember is that while all this work was being performed by certain employees, other personnel were picking up the slack for those who were working on the filters. It was probably the hardest these men had worked for the City, and I did not hear one complaint.

The donated pallet jacks. These were provided by [insert name] and were a vital part of the renovation operation.

[Insert name] expertise in innovative ideas, and turning those ideas into fabricated solutions.

In closing:

This may not sound like a gigantic accomplishment to some, but I feel this project evidences the direction we are moving in at the facilities (The most cost effective solutions for the issues we face.) Sure, it's easy to turn all these issues over to an Engineering Group for "turnkey" results. Some issues require it. But what satisfaction is there in paying top dollar for results which facility staff are capable of producing much cheaper in-house?

Thank you,

Superintendent, WWTF

EXHIBIT 9.2

City of Robinson

ROBINSON, ILLINOIS 62454

Dear Mayor and Council:

Enclosed please find my **Monthly Report** for September, 1996. This report contains information on the following:

1.0 Effluent Quality
2.0 Facility Operations
3.0 Preventive and Unscheduled Maintenance
4.0 Facility Safety and Training
5.0 Odor Complaints
6.0 Land Application of Sludge

If you should have any question concerning the Wastewater Treatment Facilities, or any other aspect of our operation, please give me a call.

Sincerely,

Superintendent, WWTF

1.0 EFFLUENT QUALITY

During September, effluent quality at the Robinson Wastewater Treatment Facility was within the limits set forth in the Illinois Environmental Protection Agency's (IEPA) NPDES permit. Table 1.0 contains effluent quality data and the respective NPDES permit limits. A copy of the Discharge Monitoring Report sent to IEPA each month is contained as Attachment A.

Table 1.0
MONTHLY EFFLUENT QUALITY DATA

Parameters	Monthly Average mg/l	IEPA Permit Limits mg/l
Ammonia Nitrogen (NH3-N)	0.12	1.5
Biochemical Oxygen Demand (CBOD5)	1.0	10.0
Total Suspended Solids (TSS)	1.4	12.0

2.0 FACILITY OPERATIONS

Operations in September were very difficult. The East Contact Stabilization Reactor completely failed on 9/6/96. The prior month it had became upset due to loadings consistently above design. Although both Reactors experienced similar problems, the West Reactor did recover. The facility staff made an all out effort to save the process, but to no avail. The tank was drained to the stormwater holding lagoon, reseeded from the parallel tank, and producing quality effluent in 2 days. **We did not** violate our NPDES discharge permit during this time.

3.0 PREVENTIVE AND UNSCHEDULED MAINTENANCE

September's scheduled maintenance for specified equipment was completed and documented.

4.0 FACILITY SAFETY AND TRAINING

We have worked 489 days without a lost-time accident. We will continue to strive for a safe workplace in order to maintain a zero-lost-time rating. On September 06, I performed the Monthly Safety Inspection of facility operations. **The inspection resulted in a 100% rating for the month.** A copy of the safety checklist is on file at the WWTF.

5.0 ODOR COMPLAINTS

There were no odor complaints in September.

6.0 LAND APPLICATION OF SLUDGE

We did not apply sludge in September.

ATTACHMENTS
A - Discharge Monitoring Report
B - Purchase Summary Report

EXHIBIT 9.2

Attachment A
Discharge Monitoring Report

PERMITTEE NAME/ADDRESS (Include Facility Name/Location if different)
NAME ROBINSON STP, CITY OF
ADDRESS 300 SOUTH LINCOLN
ROBINSON IL 62454
FACILITY ROBINSON STP, CITY OF
LOCATION ROBINSON IL 62454
ATTN: LAWRENCE E. QUICK

NATIONAL POLLUTANT DISCHARGE ELIMINATION SYSTEM (NPDES)
DISCHARGE MONITORING REPORT (DMR)
(2-16) IL0030732 PERMIT NUMBER
(17-19) 001 0 DISCHARGE NUMBER

MONITORING PERIOD
FROM YEAR 96 MO 09 DAY 01 TO YEAR 96 MO 09 DAY 30
(20-21) (22-23) (24-25) (26-27) (28-29) (30-31)

STP OUTFALL
(SUBR 04)
F - FINAL
MAJOR
EFFLUE
Form Approved. 12345
OMB No. 2040-0004
Approval expires 10-31-94
*** NO DISCHARGE |__| ***
NOTE: Read instructions before completing this form.

PARAMETER (32-37)		QUANTITY OR LOADING (3 Card Only) (46-53) AVERAGE	(54-61) MAXIMUM	UNITS	QUALITY OR CONCENTRATION (4 Card Only) (38-45) MINIMUM	(46-53) AVERAGE	(54-61) MAXIMUM	UNITS	NO. EX (62-63)	FREQUENCY OF ANALYSIS (64-68)	SAMPLE TYPE (69-70)
PH	SAMPLE MEASUREMENT	******	******		7.87	******	8.28	(12)	0	250	GR
00400 1 0 0 EFFLUENT GROSS VALUE	PERMIT REQUIREMENT	******	****** ****	****	6.0 MINIMUM	******	9.0 MAXIMUM	SU		3 DAYS WEEK	GRAB
SOLIDS, TOTAL SUSPENDED	SAMPLE MEASUREMENT	9	33	(26)	******	1.42	5.0	(19)	0	250	DC
00530 1 0 0 EFFLUENT GROSS VALUE	PERMIT REQUIREMENT	375 MO AVG	750 DAILY MX	LBS/DY	******	12 MO AVG	24 DAILY MX	MG/L		3 DAYS WEEK	COMPOS
NITROGEN, AMMONIA TOTAL (AS N)	SAMPLE MEASUREMENT	******	******	(26)	******	0.12	0.85	(19)	0	250	DC
00610 1 2 1 EFFLUENT GROSS VALUE	PERMIT REQUIREMENT	******	******	LBS/DY	******	1.5 MO AVG	3.0 DAILY MX	MG/L		3 DAYS WEEK	COMPOS
FLOW, IN CONDUIT OR THRU TREATMENT PLANT	SAMPLE MEASUREMENT	0.8	1.7	(03)	******	******	******		0	105	CN
50050 1 0 0 EFFLUENT GROSS VALUE	PERMIT REQUIREMENT	REPORT MO AVG	REPORT DAILY MX	MGD	******	******	****** ****	****		CONTINUOUS	RCOTOT
CHLORINE, TOTAL RESIDUAL	SAMPLE MEASUREMENT	******	******	(26)	******	******		(19)		500	
50060 1 0 0 EFFLUENT GROSS VALUE	PERMIT REQUIREMENT	******	******	LBS/DY	******	******	OPTIONAL DAILY MX	MG/L		CHLRNT /OCCUR	GRAB
BOD, CARBONACEOUS 05 DAY, 20C	SAMPLE MEASUREMENT	7	8	(26)	******	1.0	1.0	(19)	0	250	DC
80082 1 0 0 EFFLUENT GROSS VALUE	PERMIT REQUIREMENT	312 MO AVG	625 DAILY MX	LBS/DY	******	10 MO AVG	20 DAILY MX	MG/L		3 DAYS WEEK	COMPOS
	SAMPLE MEASUREMENT										
	PERMIT REQUIREMENT										

NAME/TITLE PRINCIPAL EXECUTIVE OFFICER
Lawrence E. Quick
Superintendent
TYPED OR PRINTED

I CERTIFY UNDER PENALTY OF LAW THAT I HAVE PERSONALLY EXAMINED AND AM FAMILIAR WITH THE INFORMATION SUBMITTED HEREIN AND BASED ON MY INQUIRY OF THOSE INDIVIDUALS IMMEDIATELY RESPONSIBLE FOR OBTAINING THE INFORMATION, I BELIEVE THE SUBMITTED INFORMATION IS TRUE, ACCURATE AND COMPLETE. I AM AWARE THAT THERE ARE SIGNIFICANT PENALTIES FOR SUBMITTING FALSE INFORMATION, INCLUDING THE POSSIBILITY OF FINE AND IMPRISONMENT. SEE 18 U.S.C. § 1001 AND 33 U.S.C. § 1319. (Penalties under these statutes may include fines up to $10,000 and or maximum imprisonment of between 6 months and 5 years.)

Lawrence E. Quick
SIGNATURE OF PRINCIPAL EXECUTIVE OFFICER OR AUTHORIZED AGENT

TELEPHONE: AREA CODE 618 NUMBER 544-8110
DATE: YEAR 96 MO 10 DAY 03

COMMENT AND EXPLANATION OF ANY VIOLATIONS (Reference all attachments here)
DMF LOAD LIMITS DISPLAYED. A 0.05 MG/L TOTAL RESIDUAL CHLORINE LIMIT APPLIES WHEN CHLORINE IS USED FOR ANY PURPOSE.

EPA Form 3320-1 (Rev. 9-88) Previous editions may be used. (REPLACES EPA FORM T-40 WHICH MAY NOT BE USED.) 00037/960723-1159 PAGE 1 OF 1

PERMITTEE NAME/ADDRESS (Include Facility Name/Location if different)
NAME ROBINSON STP, CITY OF
ADDRESS 300 SOUTH LINCOLN
ROBINSON IL 62454

FACILITY ROBINSON STP, CITY OF
LOCATION ROBINSON IL 62454
ATTN: LAWRENCE E. QUICK

NATIONAL POLLUTANT DISCHARGE ELIMINATION SYSTEM (NPDES)
DISCHARGE MONITORING REPORT (DMR)

(2-16) IL0030732 — PERMIT NUMBER
(17-19) INF L — DISCHARGE NUMBER

	MONITORING PERIOD						
	YEAR	MO	DAY		YEAR	MO	DAY
FROM	96	09	01	TO	96	09	30
	(20-21)	(22-23)	(24-25)		(26-27)	(28-29)	(30-31)

INFLUENT MONITORING
(SUBR 04) Form Approved. 12345
F - FINAL OMB No. 2040-0004
MAJOR Approval expires 10-31-94
INFLUE
*** NO DISCHARGE !__! ***
NOTE: Read instructions before completing this form.

PARAMETER (32-37)		QUANTITY OR LOADING (3 Card Only) (46-53) AVERAGE	(54-61) MAXIMUM	UNITS	QUALITY OR CONCENTRATION (4 Card Only) (38-45) MINIMUM	(46-53) AVERAGE	(54-61) MAXIMUM	UNITS	NO. EX (62-63)	FREQUENCY OF ANALYSIS (64-68)	SAMPLE TYPE (69-70)
BOD, 5-DAY (20 DEG. C)	SAMPLE MEASUREMENT	******	******	(26)	******	350	******	(19)	0	250	DC
00310 G 0 0 RAW SEW/INFLUENT	PERMIT REQUIREMENT	******	******	LBS/DY	******	REPORT MO AVG	******	MG/L		3 DAYS WEEK	COMPOS
SOLIDS, TOTAL SUSPENDED	SAMPLE MEASUREMENT	******	******	(26)	******	219	******	(19)	0	250	DC
00530 G 0 0 RAW SEW/INFLUENT	PERMIT REQUIREMENT	******	******	LBS/DY	******	REPORT MO AVG	******	MG/L		3 DAYS WEEK	COMPOS
FLOW, IN CONDUIT OR THRU TREATMENT PLANT	SAMPLE MEASUREMENT	1.10	2.03	(03)	******	******	******		0	105	CN
50050 G 0 0 RAW SEW/INFLUENT	PERMIT REQUIREMENT	REPORT MO AVG	REPORT DAILY MX	MGD	******	******	******	**** ****		CONTINUOUS	COTOT
	SAMPLE MEASUREMENT										
	PERMIT REQUIREMENT										
	SAMPLE MEASUREMENT										
	PERMIT REQUIREMENT										
	SAMPLE MEASUREMENT										
	PERMIT REQUIREMENT										
	SAMPLE MEASUREMENT										
	PERMIT REQUIREMENT										

NAME/TITLE PRINCIPAL EXECUTIVE OFFICER
Lawrence E. Quick
Superintendent
TYPED OR PRINTED

I CERTIFY UNDER PENALTY OF LAW THAT I HAVE PERSONALLY EXAMINED AND AM FAMILIAR WITH THE INFORMATION SUBMITTED HEREIN; AND BASED ON MY INQUIRY OF THOSE INDIVIDUALS IMMEDIATELY RESPONSIBLE FOR OBTAINING THE INFORMATION, I BELIEVE THE SUBMITTED INFORMATION IS TRUE, ACCURATE AND COMPLETE. I AM AWARE THAT THERE ARE SIGNIFICANT PENALTIES FOR SUBMITTING FALSE INFORMATION, INCLUDING THE POSSIBILITY OF FINE AND IMPRISONMENT. SEE 18 U.S.C. § 1001 AND 33 U.S.C. § 1319. (Penalties under these statutes may include fines up to $10,000 and or maximum imprisonment of between 6 months and 5 years.)

Lawrence E. Quick
SIGNATURE OF PRINCIPAL EXECUTIVE OFFICER OR AUTHORIZED AGENT

TELEPHONE		DATE		
618	544-8110	96	10	03
AREA CODE	NUMBER	YEAR	MO	DAY

COMMENT AND EXPLANATION OF ANY VIOLATIONS (Reference all attachments here)

EPA Form 3320-1 (Rev. 9-88) Previous editions may be used. (REPLACES EPA FORM T-40 WHICH MAY NOT BE USED.) 00001/960723-1159 PAGE 1 OF 1

PERMITTEE NAME/ADDRESS (Include Facility Name/Location if different)
NAME ROBINSON STP, CITY OF
ADDRESS 300 SOUTH LINCOLN
ROBINSON IL 62454
FACILITY ROBINSON STP, CITY OF
LOCATION ROBINSON IL 62454
ATTN: LAWRENCE E. QUICK

NATIONAL POLLUTANT DISCHARGE ELIMINATION SYSTEM (NPDES)
DISCHARGE MONITORING REPORT (DMR)

(2-16) IL0030732 PERMIT NUMBER — (17-19) STR M DISCHARGE NUMBER

IN-STREAM MONITORING
(SUBR 04)
F - FINAL
MAJOR
STREAM
*** NO DISCHARGE | __ | ***

Form Approved. OMB No. 2040-0004 Approval expires 10-31-94 — 12345

	MONITORING PERIOD						
	YEAR	MO	DAY		YEAR	MO	DAY
FROM	96	09	01	TO	96	09	30
	(20-21)	(22-23)	(24-25)		(26-27)	(28-29)	(30-31)

NOTE: Read instructions before completing this form.

PARAMETER (32-37)		(3 Card Only) QUANTITY OR LOADING (46-53) AVERAGE	(54-61) MAXIMUM	UNITS	(4 Card Only) QUALITY OR CONCENTRATION (38-45) MINIMUM	(46-53) AVERAGE	(54-61) MAXIMUM	UNITS	NO. EX (62-63)	FREQUENCY OF ANALYSIS (64-68)	SAMPLE TYPE (69-70)
TEMPERATURE, WATER DEG. FAHRENHEIT	SAMPLE MEASUREMENT	******	******		******	******	76	(15)	0	240	GR
00011 O 0 0 SEE COMMENTS BELOW	PERMIT REQUIREMENT	******	******	**** ****	******	******	REPORT DAILY MX	DEG.F		WEEKLY	
TEMPERATURE, WATER DEG. FAHRENHEIT	SAMPLE MEASUREMENT	******	******		******	******	74	(15)	0	240	GR
00011 P 0 0 SEE COMMENTS BELOW	PERMIT REQUIREMENT	******	******	**** ****	******	******	REPORT DAILY MX	DEG.F		WEEKLY	
TEMPERATURE, WATER DEG. FAHRENHEIT	SAMPLE MEASUREMENT	******	******		******	******	66	(15)	0	240	GR
00011 Q 0 0 SEE COMMENTS BELOW	PERMIT REQUIREMENT	******	******	**** ****	******	******	REPORT DAILY MX	DEG.F		WEEKLY	
TEMPERATURE, WATER DEG. FAHRENHEIT	SAMPLE MEASUREMENT	******	******		******	******	68	(15)	0	240	GR
00011 R 0 0 SEE COMMENTS BELOW	PERMIT REQUIREMENT	******	******	**** ****	******	******	REPORT DAILY MX	DEG.F		WEEKLY	
TEMPERATURE, WATER DEG. FAHRENHEIT	SAMPLE MEASUREMENT	******	******		******	******	70	(15)	0	240	GR
00011 S 0 0 SEE COMMENTS BELOW	PERMIT REQUIREMENT	******	******	**** ****	******	******	REPORT DAILY MX	DEG.F		WEEKLY	
PH	SAMPLE MEASUREMENT	******	******		******	******	8.20	(12)	0	240	GR
00400 O 0 0 SEE COMMENTS BELOW	PERMIT REQUIREMENT	******	******	**** ****	******	******	REPORT DAILY MX	SU		WEEKLY	
PH	SAMPLE MEASUREMENT	******	******		******	******	7.87	(12)	0	240	GR
00400 P 0 0 SEE COMMENTS BELOW	PERMIT REQUIREMENT	******	******	**** ****	******	******	REPORT DAILY MX	SU		WEEKLY	

NAME/TITLE PRINCIPAL EXECUTIVE OFFICER: Lawrence E. Quick, Superintendent (TYPED OR PRINTED)

I CERTIFY UNDER PENALTY OF LAW THAT I HAVE PERSONALLY EXAMINED AND AM FAMILIAR WITH THE INFORMATION SUBMITTED HEREIN; AND BASED ON MY INQUIRY OF THOSE INDIVIDUALS IMMEDIATELY RESPONSIBLE FOR OBTAINING THE INFORMATION, I BELIEVE THE SUBMITTED INFORMATION IS TRUE, ACCURATE AND COMPLETE. I AM AWARE THAT THERE ARE SIGNIFICANT PENALTIES FOR SUBMITTING FALSE INFORMATION, INCLUDING THE POSSIBILITY OF FINE AND IMPRISONMENT. SEE 18 U.S.C. § 1001 AND 33 U.S.C. § 1319. (Penalties under these statutes may include fines up to $10,000 and or maximum imprisonment of between 6 months and 5 years.)

SIGNATURE OF PRINCIPAL EXECUTIVE OFFICER OR AUTHORIZED AGENT: Lawrence E. Quick

TELEPHONE		DATE		
618	544-8110	96	10	03
AREA CODE	NUMBER	YEAR	MO	DAY

COMMENT AND EXPLANATION OF ANY VIOLATIONS (Reference all attachments here)

MONITORING LOCATIONS: O=1ST WEEK, P=2ND WEEK, Q=3RD WEEK, R=4TH WEEK, S=5TH WEEK

EPA Form 3320-1 (Rev. 9-88) Previous editions may be used. (REPLACES EPA FORM T-40 WHICH MAY NOT BE USED) 00013/960723-1159 PAGE 1 OF 2

PERMITTEE NAME/ADDRESS (Include Facility Name/Location if different)
NAME ROBINSON STP, CITY OF
ADDRESS 300 SOUTH LINCOLN
ROBINSON IL 62454
FACILITY ROBINSON STP, CITY OF
LOCATION ROBINSON IL 62454
ATTN: LAWRENCE E. QUICK

NATIONAL POLLUTANT DISCHARGE ELIMINATION SYSTEM (NPDES)
DISCHARGE MONITORING REPORT (DMR)

IL0030732 — PERMIT NUMBER
STR M — DISCHARGE NUMBER

MONITORING PERIOD
FROM YEAR 96 MO 09 DAY 01 TO YEAR 96 MO 09 DAY 30

IN-STREAM MONITORING
(SUBR 04)
F - FINAL
MAJOR
STREAM
*** NO DISCHARGE |__| ***

Form Approved. 12345
OMB No. 2040-0004
Approval expires 10-31-94

NOTE: Read instructions before completing this form.

PARAMETER (32-37)		QUANTITY OR LOADING (3 Card Only) AVERAGE (46-53)	MAXIMUM (54-61)	UNITS	QUALITY OR CONCENTRATION (4 Card Only) MINIMUM (38-45)	AVERAGE (46-53)	MAXIMUM (54-61)	UNITS	NO. EX (62-63)	FREQUENCY OF ANALYSIS (64-68)	SAMPLE TYPE (69-70)
PH	SAMPLE MEASUREMENT	******	******		******	******	8.10	(12)	0	240	GR
00400 Q 0 0 SEE COMMENTS BELOW	PERMIT REQUIREMENT	******	******	**** ****	******	******	REPORT DAILY MX	SU		WEEKLY	
PH	SAMPLE MEASUREMENT	******	******		******	******	7.97	(12)	0	240	GR
00400 R 0 0 SEE COMMENTS BELOW	PERMIT REQUIREMENT	******	******	**** ****	******	******	REPORT DAILY MX	SU		WEEKLY	
PH	SAMPLE MEASUREMENT	******	******		******	******	7.90	(12)	0	240	GR
00400 S 0 0 SEE COMMENTS BELOW	PERMIT REQUIREMENT	******	******	**** ****	******	******	REPORT DAILY MX	SU		WEEKLY	
	SAMPLE MEASUREMENT										
	PERMIT REQUIREMENT										
	SAMPLE MEASUREMENT										
	PERMIT REQUIREMENT										
	SAMPLE MEASUREMENT										
	PERMIT REQUIREMENT										
	SAMPLE MEASUREMENT										
	PERMIT REQUIREMENT										

NAME/TITLE PRINCIPAL EXECUTIVE OFFICER
Lawrence E. Quick
Superintendent
TYPED OR PRINTED

I CERTIFY UNDER PENALTY OF LAW THAT I HAVE PERSONALLY EXAMINED AND AM FAMILIAR WITH THE INFORMATION SUBMITTED HEREIN; AND BASED ON MY INQUIRY OF THOSE INDIVIDUALS IMMEDIATELY RESPONSIBLE FOR OBTAINING THE INFORMATION, I BELIEVE THE SUBMITTED INFORMATION IS TRUE, ACCURATE AND COMPLETE. I AM AWARE THAT THERE ARE SIGNIFICANT PENALTIES FOR SUBMITTING FALSE INFORMATION, INCLUDING THE POSSIBILITY OF FINE AND IMPRISONMENT. SEE 18 U.S.C. § 1001 AND 33 U.S.C. § 1319. (Penalties under these statutes may include fines up to $10,000 and or maximum imprisonment of between 6 months and 5 years.)

SIGNATURE OF PRINCIPAL EXECUTIVE OFFICER OR AUTHORIZED AGENT

TELEPHONE: AREA CODE 618 NUMBER 544-8110
DATE: YEAR 96 MO 10 DAY 03

COMMENT AND EXPLANATION OF ANY VIOLATIONS (Reference all attachments here)

MONITORING LOCATIONS: O=1ST WEEK, P=2ND WEEK, Q=3RD WEEK, R=4TH WEEK, S=5TH WEEK

EPA Form 3320-1 (Rev. 9-88) Previous editions may be used. (REPLACES EPA FORM T-40 WHICH MAY NOT BE USED.) 00014/960723-1159 PAGE 2 OF 2

Attachment B
Purchase Summary Report

ROBINSON WASTEWATER TREATMENT PLANT
PURCHASE SUMMARY FOR SEPTEMBER 1996

	Cost	Vendor	Explanation
$	500.00	WW MONROE EQ. CO.	BELT PRESS CONVEYOR BELT
$	50.43	RURAL KING	MISC. SUPPLIES
$	106.00	CARTER LUMBER	SUPPLIES TO BUILD SLUDGE HOPPER
$	57.62	MECH. LAUNDRY	WEEKLY UNIFORM SERVICE
$	68.90	HACH	2 PACKS COD AMPULES
$	50.00	FISHER	NH3 MEMBRANE CAPS
$	10.18	BIG BUCK	HARDWARE FOR SLUDGE CONVEYOR
$	58.28	MECH. LAUNDRY	WEEKLY UNIFORM SERVICE
$	53.05	AMERITECH	CELLULAR BILL
$	80.44	RURAL KING	PAINT & SUPPLIES FOR PRESS ROOM FLOOR
$	24.12	RURAL KING	ROPE, HOOKS & SNAPS FOR SLUDGE HOPPER
$	12.68	CARTER LUMBER	SUPPLIES FOR HOPPER
$	7.30	BIG BUCK	SAW BLADE
$	3.20	VANDEVANTER	8 O-RINGS
$	22.90	HACH	NUTRIENT BUFFER PILLOWS
$	5.40	AT&T	MAINT. L.D. BILL
$	24.28	RURAL KING	RIVETS, OXYGEN & SEALANT
$	9.22	SERVICE STAR	SEND BACK SHAFT SLEEVES
$	56.96	MECH. LAUNDRY	WEEKLY UNIFORM SERVICE
$	1.59	RURAL KING	BOLTS & ELECT. TAPE
$	142.40	E.C. LABS	1/4LY SLUDGE ANALYSIS
$	7.00	MARATHON TIRE	REPAIR JEEP TIRE
$	1.33	AT&T	WIRELESS LONG DISTANCE
$	450.00	O-BRIAN	ROOT CUTTER
$	378.51	O-BRIAN	PENETRATOR NOZZLES
$	481.20	W.W. MONROE	BELT FOR CONVEYOR

$ 70.30	COMMUNICATION PRODUCTS	INSTALL RADIO FROM CHEVY TRUCK IN TAURUS
$ 74.86	RURAL KING	BATTERIES, BOOTS & RAIN SUIT FOR MONTY
$ 24.00	IGA	24 GALLON DISTILLED H2O
$ 42.17	GTE	LAB PHONE BILL
$ 43.02	GRAINGER	SAFETY GLASSES
$ 56.96	MECH. LAUNDRY	WEEKLY UNIFORM SERVICE
$ 35.00	YSI	BOD STIRRING ROD
$ 129.38	CRAWFORD HYDRAULICS	16' OF 1" LEADER HOSE FOR JET TRUCK
$ 14.49	NAPA	BATTERY CABLE PARTS
$ 37.16	RURAL KING	GUN GREASE & PICK HANDLES
$ 63.00	R&L CONCRETE	3/4 YD FOR REPAIR OF SEWER IN STREET
$ 29.98	NAPA	CHOKE CABLE FOR TRUCK
$ 36.65	BRADFORD	13' - 8" PVC PIPE
$ 53.94	MCMASTER CARR	2 WIRE BRISTLE BROOMS
$ 94.42	GLOBAL SAFETY	EMERGENCY LIGHTS
$ 18.49	NAPA	DRILL BIT & BEARING
$ 75.67	RURAL KING	PARTS FOR 20 HP MOWER
$ 56.30	MECH. LAUNDRY	WEEKLY UNIFORM SERVICE
$ 151.00	MICROFLEX	2 CS LATEX GLOVES
$ 74.00	R&L CONCRETE	CC FOR STREET REPAIR.
$ 39.93	BUEHLERS	LIQUID SOAP
$ 8.18	NAPA	DRILL BIT & PUNCH
$ 18.00	GRAVES AUTO	3000 MILE MAINTENANCE ON FORD TAURUS.

$ 3,909.89 SEPTEMBER, 1996 Total

EXHIBIT 9.3

City of Robinson Mayor and City Council
Robinson, Illinois 62454

Dear Officials:

Enclosed please find my "Annual Report" for the year starting [date] to [date]. This Annual Report contains information on the following:

1.0 Effluent Quality
2.0 Facility Operations
3.0 Preventive and Unscheduled Maintenance
4.0 Facility Safety and Training
5.0 Land Application of Sludge
6.0 Other

Please feel free to contact me if should you have questions about this report or any other aspect of our operation.

Sincerely,

CITY OF ROBINSON

Superintendent, WWTF

1.0 Effluent Quality

The Robinson Wastewater Treatment Facility effluent quality was within the Illinois Environmental Protection Agency's (IEPA) NPDES permit limits during the reporting period. The year's Effluent Quality Data is contained as Attachment A. Influent loading data for the period are contained as Attachment B. Graphs showing levels of the two major effluent parameters are contained as Attachment C.

2.0 Facility Operations

Listed below are some of the projects which were completed in this reporting period.

- Completed all necessary Pretreatment Program tasks. (Currently awaiting USEPA approval)

- Via the Pretreatment Program, SIU's have mitigated concentrations of individual troublesome constituents which cause the City's WWTF problems.
- Completed Fine Bubble Diffuser retrofit.
- Completed Pickford Street Lift Station upgrades.
- Currently in the process of starting Sludge Storage Building.
- Developed, submitted and received approval of "Plan for Toxicity Reduction Evaluation." Although this plan took much time to put together, I was happy it was not needed due to impressive results of the following two analyses.
- Passed all 6 Biomonitoring events on effluent.
- Passed all 6 metals analyses on effluent.
- Complied with all USEPA 503 sludge regulations.
- Settled the enormous issue of WWTP BOD loadings. (Industry building a pretreatment plant)
- Reduced electrical costs due to new diffuser system. (Electrical costs will drop even more after the industry is pretreating.)
- Reduced potable water consumption with better quality digested sludge due to Fine Bubble Diffusers. (Reduced to approximately 1/3 of previous)
- Identified and required correction of significant Infiltration/Inflow points in the collection system.
- Removed leaking valve in lagoon influent line over creek.
- Reduced polymer costs drastically due to better quality digested sludge. (Fine Bubble Diffusers)
- Reduced WWTP actual average influent flow values by aggressive flow management techniques. (Was 1.4 MGD, now 1.1 MGD)

Although this is a partial list of accomplishments, I believe it continues to evidence our commitment to provide the most cost-effective wastewater treatment for the City of Robinson.

Note: Most of the above listed tasks were performed by facility personnel.

3.0 Preventive and Unscheduled Maintenance

Scheduled maintenance for specified equipment was completed. All facility equipment was maintained and lubricated per manufacturers specifications and entered in the computerized maintenance database.

4.0 Facility Safety and Training

Since April 4, 1995, we have had one lost-time accident. The number of no-lost-time days since the accident is 310.

We have worked hard to build up and maintain the facility safety rating. The safety rating for the facility has been at 100% for 9 consecutive months. Monthly safety inspections at the facility are performed to evaluate progress.

We continue to provide monthly safety seminars to WWTF and Street Department employees. Fourteen topics are covered in the seminars held each year.

Seminars/Training

- Electrical Safety
- Laboratory Safety
- Confined Space Entry & Policy
- Prevention of Back Injury
- Superfund Amendment Reauthorization Act (SARA)
- Vehicle Safety
- Fire Safety and Prevention
- Trenching Safety
- Facility Safety Equipment
- Review Safety Rules
- Earthquake
- Emergency Response Plan Review
- Bloodborne Pathogens
- First-Aid Kit Training

5.0 Land Application of Sludge

We applied approximately 243 dry tons of sludge to local farm ground between April 4, 1995 and April 3, 1996. All sludge was applied in accordance with IEPA and Federal EPA permit guidelines.

6.0 Other

We provided a 14-week advanced wastewater treatment course to five WWTF employees at Lake Land College, numerous seminars and conferences. We continue training to further upgrade licences. Our ultimate goal is for all WWTF employees to be Illinois Class 1 operators by the year 2000.

Attachments

A. Effluent Quality
B. Influent Loading
C. Influent and Effluent Quality Graphs

Attachment A
Robinson WWTF
Effluent Quality Data
Monthly Average
mg/l

MONTH	CBOD5 mg/l	TSS mg/l	NH3-N mg/l
APRIL 1995	3	2	0.08
MAY	2	1	0.13
JUNE	3	2	0.40
JULY	1	1	0.20
AUGUST	2	1	0.18
SEPTEMBER	2	2	0.13
OCTOBER	2	1	0.22
NOVEMBER	1	2	0.27
DECEMBER	2	2	0.33
JANUARY 1996	4	4	0.42
FEBRUARY	4	3	0.51
MARCH	4	3	0.65
YEARLY AVERAGE	2.5	2.0	0.29

IEPA Limits:

CBOD5: 10 mg/l

TSS: 12 mg/l

NH3-N: April - October (1.5 mg/l)
November - March (1.9 mg/l)

Attachment B
Robinson WWTF
Influent Loading Data
Monthly Average
Pounds Per Day

MONTH	BOD5 LBS/DAY	TSS LBS/DAY	INFLUENT FLOW MGD
APRIL 95	2565	2430	1.6
MAY	2995	3012	2.1
JUNE	3874	4070	1.5
JULY	2645	1962	1.3
AUGUST	2309	1959	1.3
SEPTEMBER	2302	2137	1.1
OCTOBER	1853	1760	1.0
NOVEMBER	1982	1709	1.1
DECEMBER	2506	1780	1.2
JAN 96	2123	2630	1.9
FEBRUARY	3182	3303	1.6
MARCH	2451	3426	1.7
YEARLY AVERAGE	2566	2515	1.5

ROBINSON WWTP AVERAGE DESIGN CAPACITY:

BOD5 LBS/DAY: 2550

TSS LBS/DAY : 3300

FLOW MGD : 1.5

Attachment C
City of Robinson WWTF
Major Effluent Quality Parameters

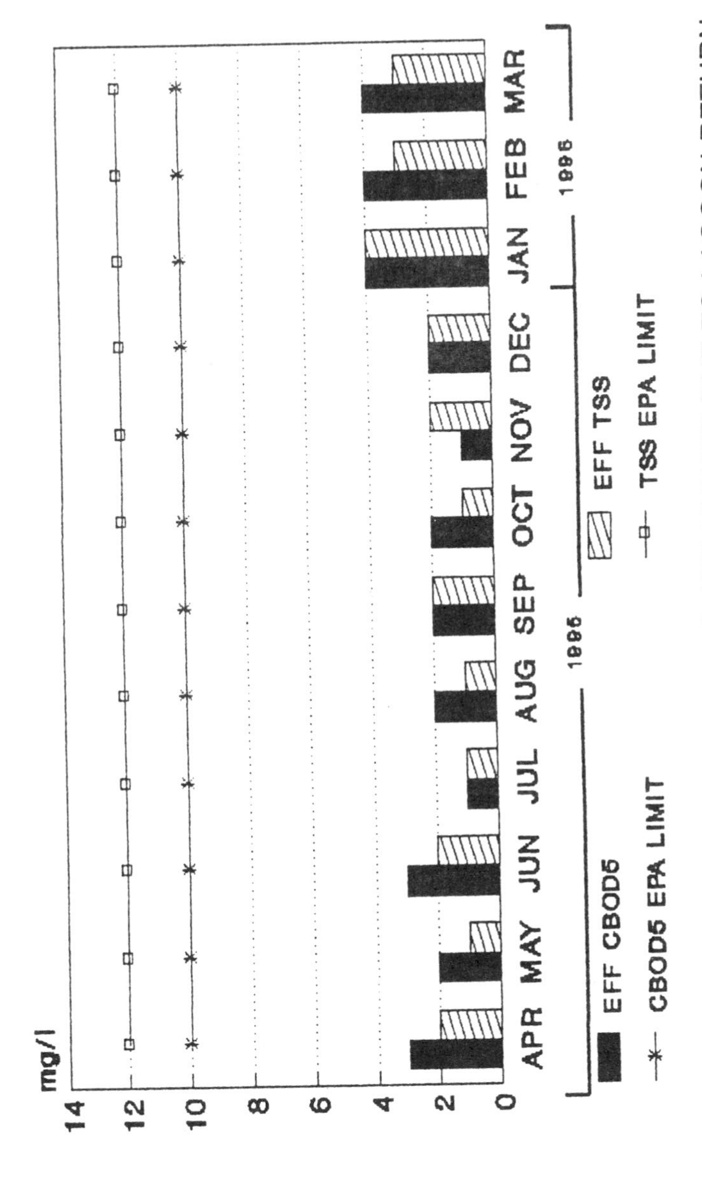

NOTE: JAN, FEB, AND MAR VALUES ARE HIGHER DUE TO LAGOON RETURN
MONTHLY AVERAGES (mg/l)

CHAPTER 10

Equipment Start-up

Once the equipment is installed, maintenance on it, as well as all other process equipment in the facility, is vital. You should read through the operations and maintenance manuals and familiarize yourself with the specifics of the equipment. Question the following:

- How often does the oil need to be changed and what type of oil?
- Where are the grease lubrication points and what type of grease?
- What spare parts should I keep in stock?
- Where are the adjustments?

If you do not understand any aspect of the equipment, ask the equipment representative and/or manufacturer. The equipment may contain "break-in oil," which should be changed after the initial 60 days of operation. These types of information are very important because it may drastically affect the length of service life of the equipment. Also, equipment warranties, which can be lifesavers if premature failure occurs, could be voided if it was not properly maintained.

I have found the best mechanism to track and document

equipment maintenance is through the use of a personal computer containing maintenance management software. You may think of these as expensive tools, but they really are not. A good simple system would include a personal computer at about $1,200.00, a printer at about $300.00, and software as inexpensive as $1,000.00. So for under $3,000.00 you could have a computerized maintenance system. How many times have you set up a written maintenance schedule and after a few months it falls by the wayside? Equipment maintenance frequencies are forgotten, and nobody has any idea where you are. Computerized systems create an atmosphere of responsibility, ensuring tasks are performed. Work orders generated by the software stay on the computer until they are closed out. So if you issue your maintenance people 10 work orders that are due to be performed that week, they perform the tasks and turn the completed forms back in. This creates an atmosphere of accountability. Of course a computer cannot change oil in a gear reducer or watch to see if an employee actually performs the work. Therefore, checks may need to be performed from time to time ensuring the work has been performed. I have found that once the maintenance staff realizes the reactive maintenance duties (emergency repairs) are minimized from good scheduled maintenance, they are more than willing to perform the scheduled tasks.

Depending on what software you choose, the basic minimum capabilities should be as follows:

- equipment name entry (Example, Figure 10.1)
- equipment task entry
- spare parts inventory (Example, Figure 10.2)
- work order generation
- work order history reports (Example, Figure 10.3)
- scheduled/repair maintenance entry (Example, Figure 10.4)
- work order close entry (Example, Figure 10.5)
- equipment/task/work order lookup tables
- graph generation (Example, Figure 10.6)

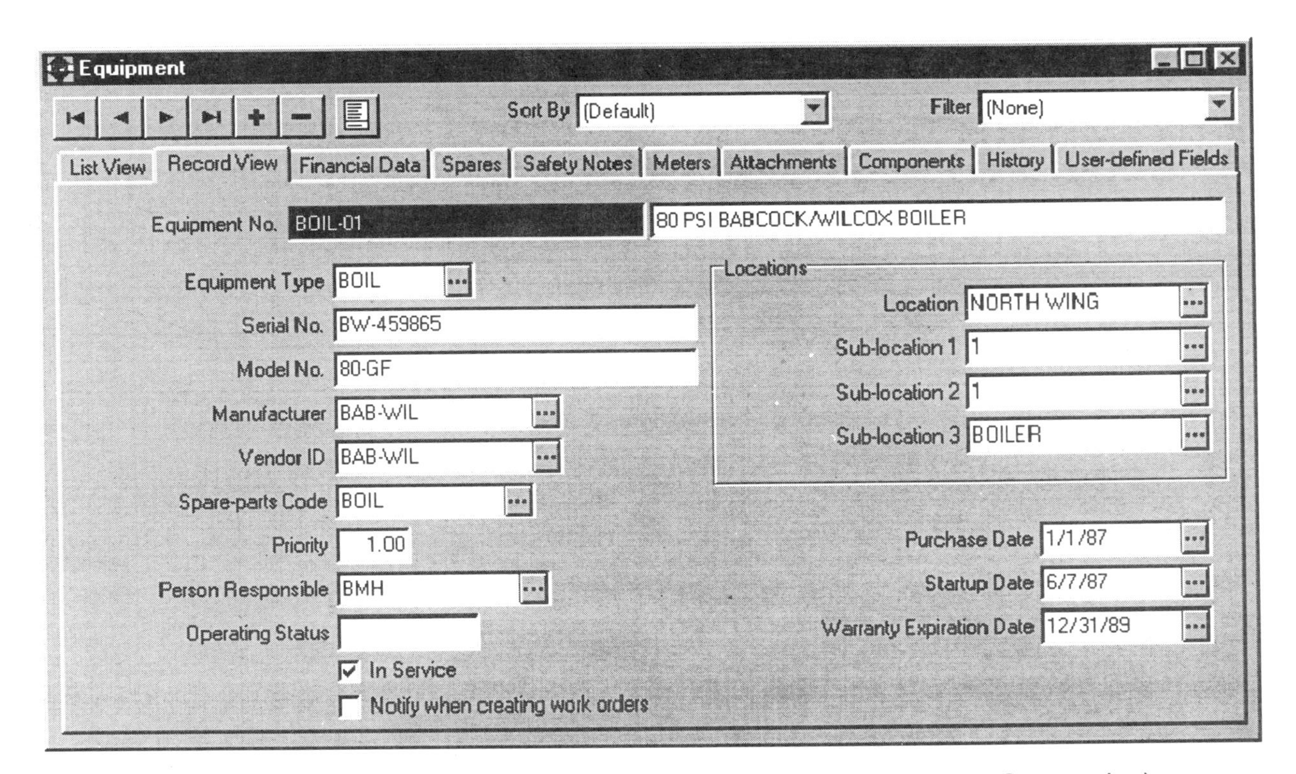

Figure 10.1. Equipment entry. (Reproduced with permission from Datastream Systems, Inc.)

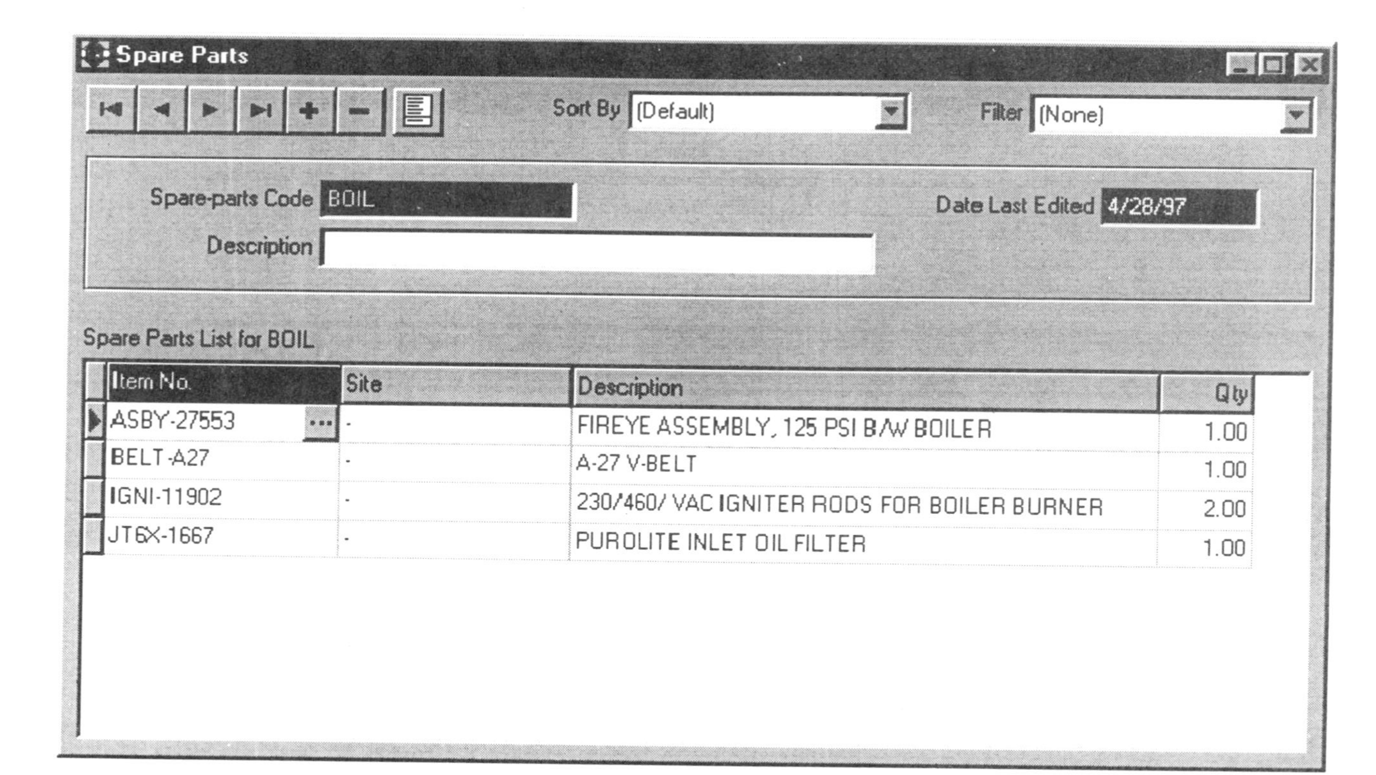

Figure 10.2. Spare parts inventory. (Reproduced with permission from Datastream Systems, Inc.)

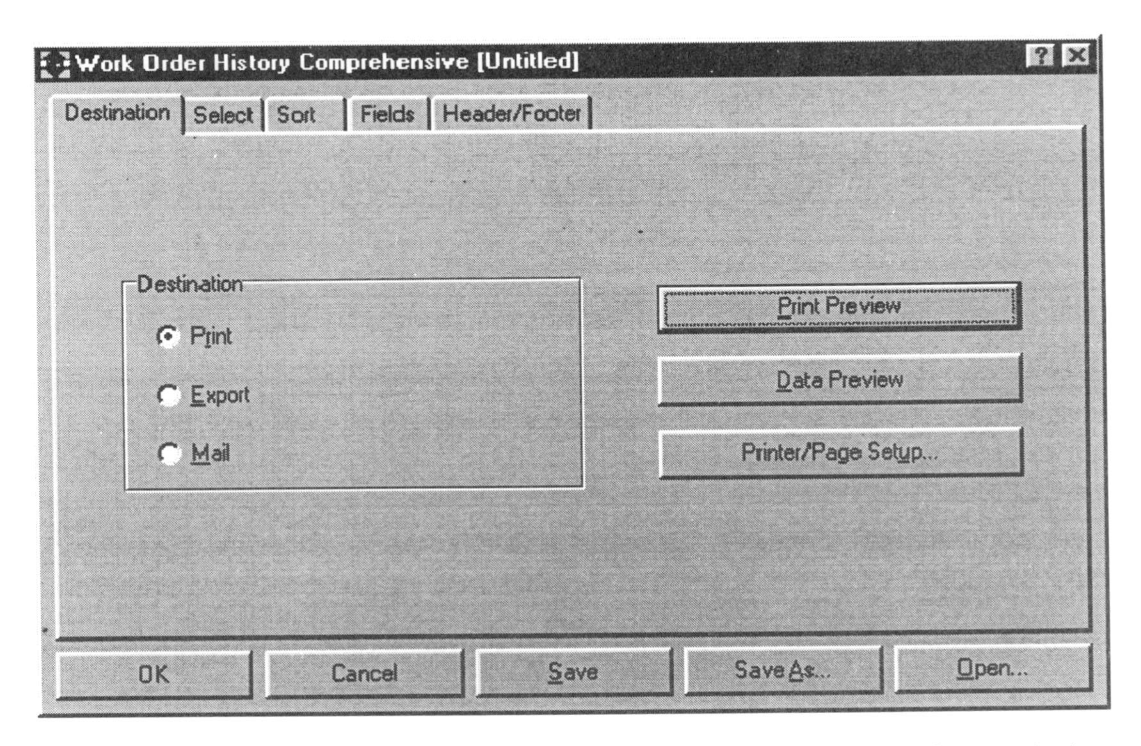

Figure 10.3. Work order history. (Reproduced with permission from Datastream Systems, Inc.)

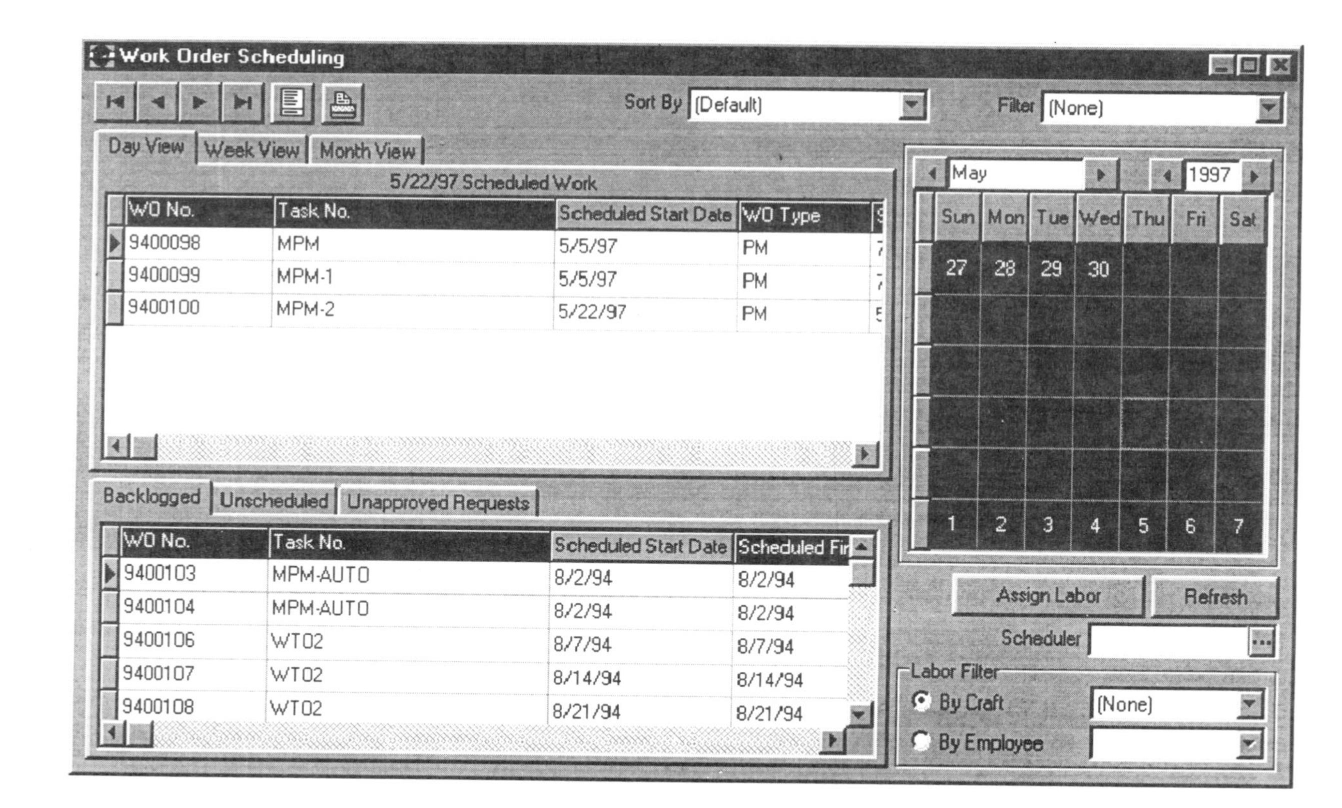

Figure 10.4. Scheduling repairs/maintenance. (Reproduced with permission from Datastream Systems, Inc.)

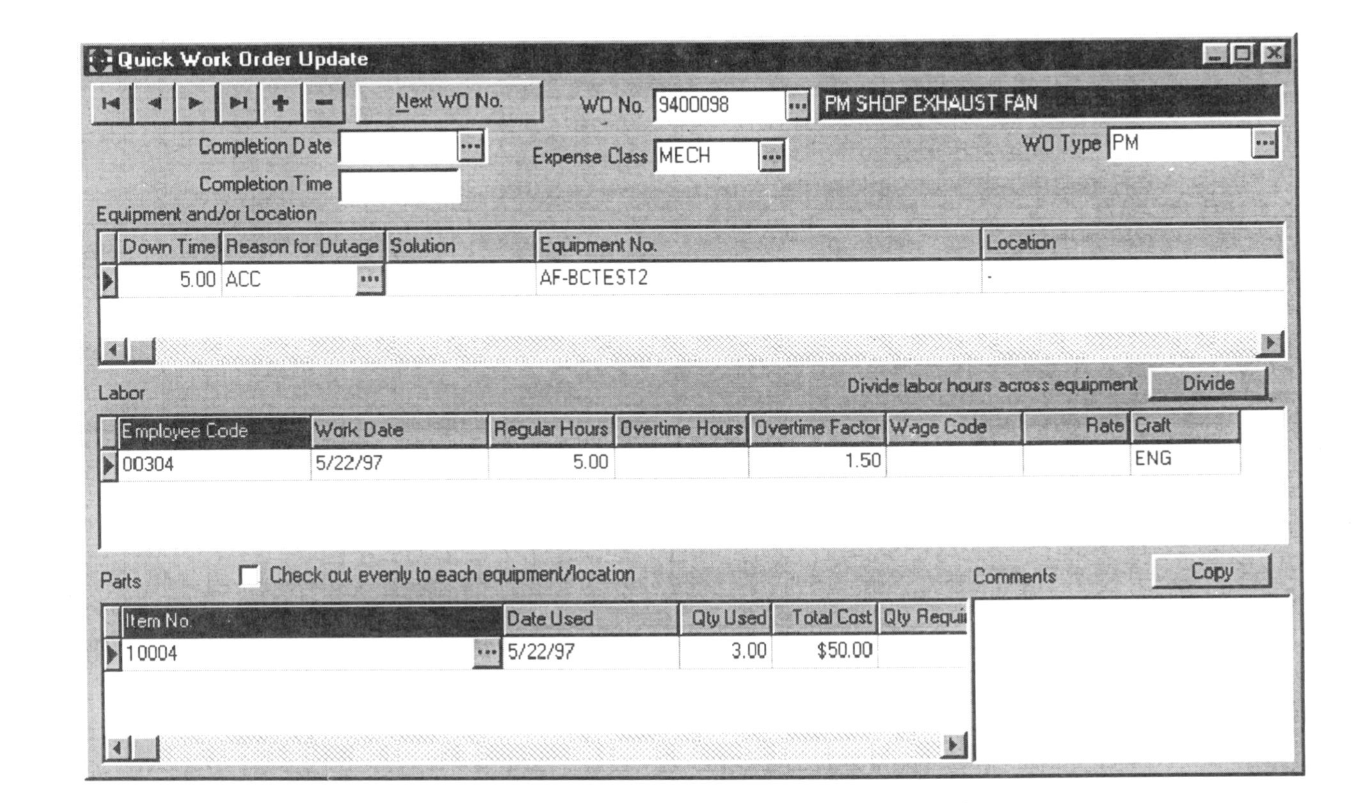

Figure 10.5. Repair/maintenance entry. (Reproduced with permission from Datastream Systems, Inc.)

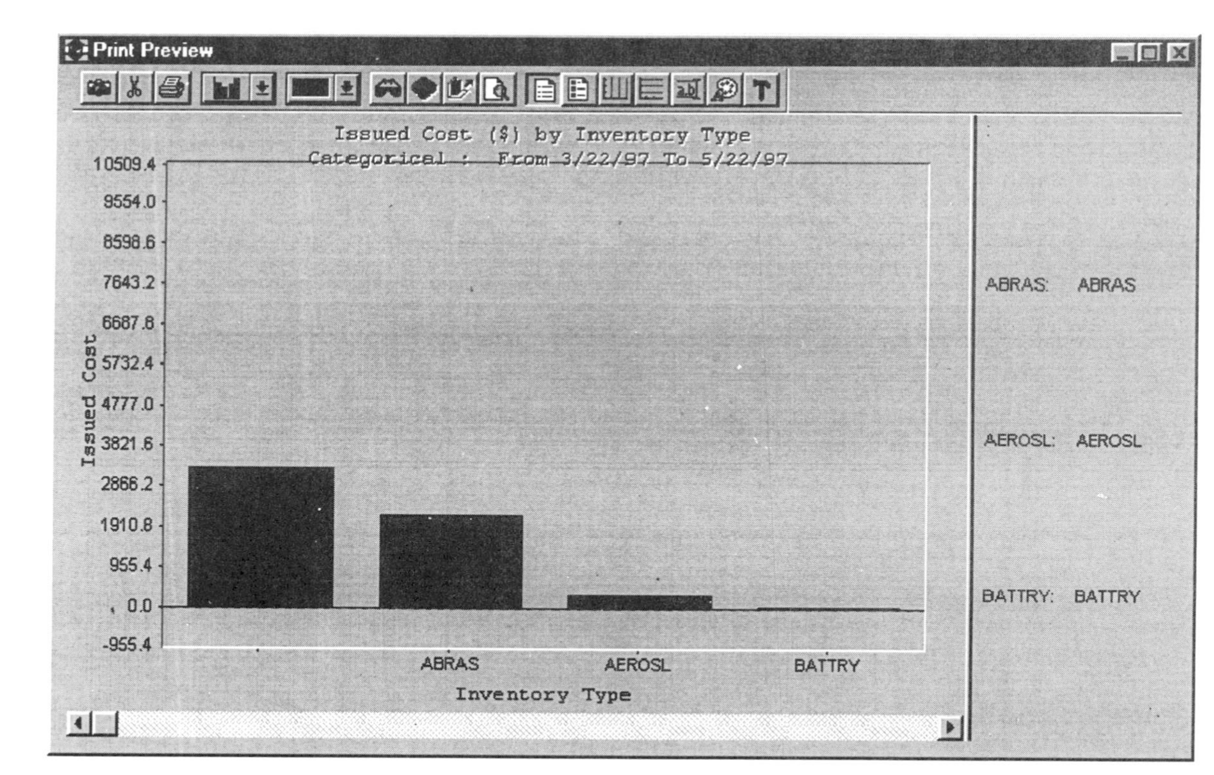

Figure 10.6. Graph generation. (Reproduced with permission from Datastream Systems, Inc.)

Due to the fact that $3,000.00 may seem outrageous to the board, good sound rationale may be needed. I have found, and I'm sure you will too, that a system such as this will save you its price in the first year. Because all manufacturer recommended tasks at required frequencies will be performed, the equipment will remain in tip-top shape. Items, such as lists of instructions, safety issues, parts, tools, lubricants needed for the task, etc., can be included on the tasks. These items will then print out on the work orders. (An example of a work order is found in Figure 10.7.)

Take, for instance, a submersible pump. This pump is to have the oil changed in it once a year. Let's assume that the seal is leaking and nobody changed the oil when it was due. If the oil had been changed, the mechanic would have noticed it had a milky color. He would also know the seal had either failed or was close to failing. Now it's necessary to send the pump to a motor repair shop for rewinding, new bearings, seals, labor for electricians and mechanics, etc. This one pump could easily cost as much as the computer hardware and software.

Here is another hypothetical situation. Let's say a bearing fails on a belt filter press. It is an identical bearing to one that was replaced 3 months ago. A repair work order was prepared for the previous bearing failure, which contained the part name, number, cost, supplier, etc. You can retrieve the information from your computer database in a matter of seconds, compared with digging for hours in manuals.

Because computers are very user-friendly nowadays, almost anyone can master them. The software is menu driven (a list of choices appears at the top of the screen to choose from, an example menu is shown in Figure 10.8) and can be self taught in a short period of time. User-friendliness should be a significant feature demanded from a software provider. They should also provide technical support either included with the package or for a fee. If you are a first time user, I would recommend

CITY OF ROBINSON
WWTF
WORK ORDER PRINT REPORT

Page 1

```
Work Order #: 40001      (SCHED      )    Task #:
CHANGE GEAR REDUCER OIL IN CLARIFIER DRIVE
=============================================================================
Equipment #...  E C/S DRIVE              Warranty Expires  2/2/2000
EQ Description  E C/S DRIVE MOTOR
Location......  WWTP
Department....  1                        Current Meter... __________
Cost Center...  C/S TANKS
-----------------------------------------------------------------------------
Originator....  SUPT                     Request Date....
Phone.........  555-555-555              Extension.......
-----------------------------------------------------------------------------
Start Date....  1/1/97                   Craft...........  MAINT
Finish Date...                           Crew Size.......  2
Priority......  1.00                     Est Labor Hours.  1.50
RFO Code......  _______ (   )
Down Time.....  _______                  Lot #...........  ______________

================================ LABOR =======================================
Employee          Craft         Name                                    Hours

_______________  ___________  ____________________________________  _______
_______________  ___________  ____________________________________  _______
_______________  ___________  ____________________________________  _______
_______________  ___________  ____________________________________  _______
_______________  ___________  ____________________________________  _______

============================ COMMENTS & NOTES ================================

Comments: ___________________________________________________________________

_____________________________________________________________________________

Notes: ______________________________________________________________________

_____________________________________________________________________________


================================= COMMENTS ===================================
LOCK AND TAG OUT EQUIPMENT BEFORE BEGINNING TASK.
CHECK TO BE SURE WE HAVE THE SPECIFIED OIL IN STOCK (AMOCO 460)
"DO NOT OVERFILL GEAR REDUCER, 4 QTS IS THE REQUIRED AMOUNT"

================================== PARTS =====================================
Item #                Qty Reqd Description
-------------------- -------- -----------------------------------------------
CYL 460                        AMOCO 460 CYLINDER OIL
```

Figure 10.7.

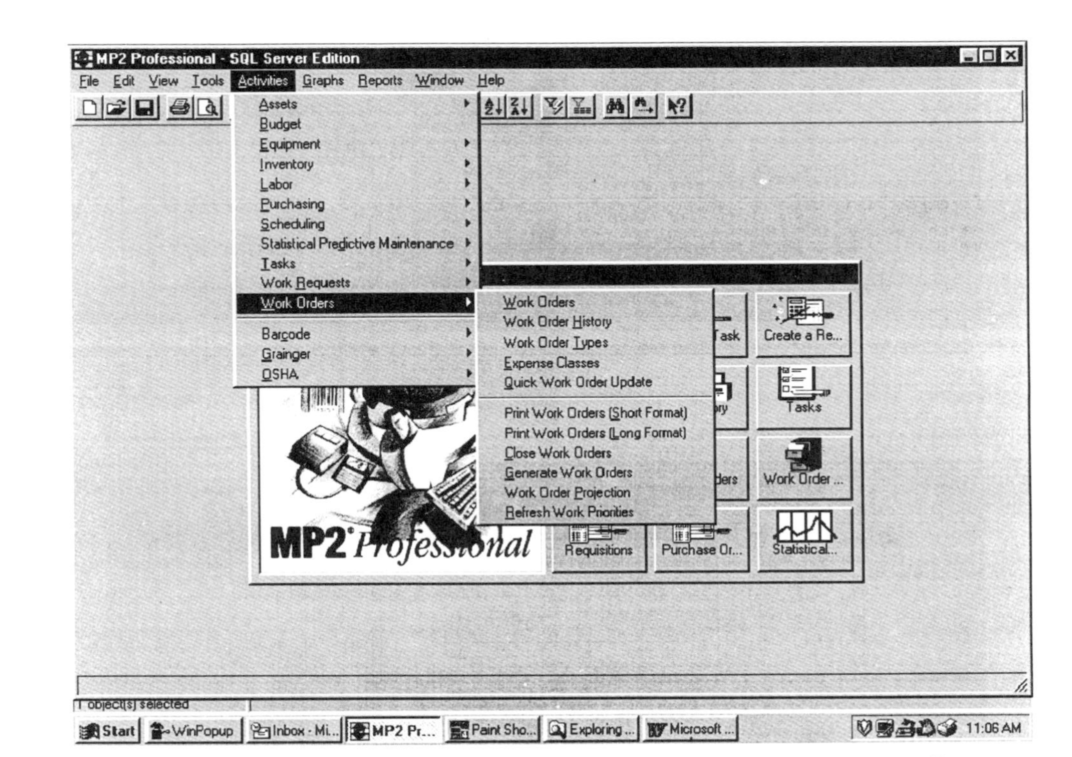

Figure 10.8. Example menu. (Reproduced with permission from Datastream Systems, Inc.)

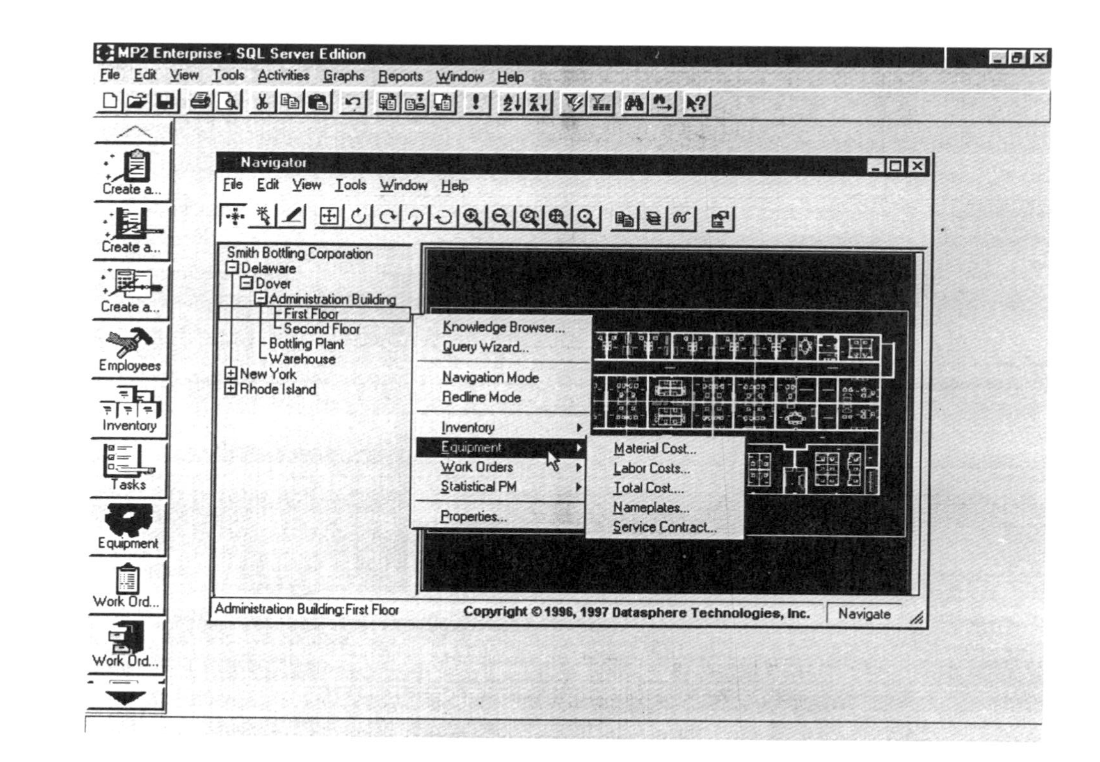

Figure 10.8 (continued). Example menu. (Reproduced with permission from Datastream Systems, Inc.)

that you get the support for at least a year or so until you feel comfortable with the program. This way if you have problems, you can call the software provider and they will take you through step by step, if necessary, to get you back up and running. *Note:* I know this because I had to use them quite a few times when I started.

Now, say justification is needed to replace a piece of failing equipment. Generate a report from the work order history database, which will contain repair data, e.g., man-hours, parts costs, labor costs, contract labor costs, expensive downtime data, etc. This could prove that it is more cost-effective to replace the equipment than to keep it. This is a very powerful tool and can help in many different ways.

Scheduled maintenance is very significant in the eyes of our regulatory agency as well. They believe, and justifiably so, that if equipment is not maintained properly, the facility will be a very good candidate for noncompliance in the future. It is so nice when going through our yearly EPA audit and being asked to provide maintenance records that I can generate them in a matter of seconds.

These procedures can be, and have been, performed by handwritten documentation very proficiently. If one chooses to perform maintenance management this way, that is fine too. However, regardless of how maintenance is performed and tracked, it is very vital it is done. Don't let it slide!!

CHAPTER 11

Similar Equipment

In the past, because municipalities typically purchased equipment based on the "low bid" concept, a potpourri of equipment may exist in your facilities. This may have been hard to avoid due to mandates to award low bid from the granting or lending authority. If you are upgrading or retrofitting equipment that has been budgeted and are going to pay for it without assistance from grants or low-interest loans, choose whichever manufacturer is best.

I have found through experience to stick with a manufacturer with a good track record. This can be beneficial in numerous ways. Take, for instance, sewage pumping stations. They historically fit the potpourri (many different brands) scenario. If you have 10 pump stations and need to retrofit new pumps in three of the stations (submersible or drypit), the same type of pump and motor may work in all three. It may be necessary to use different variations of impellers to meet the discharge curve criteria, or maybe two are drypit, and one is a submersible station. The same model of pump may work in all three. A combination we use is the drypit submersible. These are offered by many legitimate pump manufacturers. Dependent on the application, they may be installed as submerged pumps in a wetwell or in a drypit application using adaptable pump stands and volutes (Figure 11.1) that contain suction and discharge piping. The reason these are so practical is that they can be installed as drypit submersibles in an existing drywell that may be susceptible to flooding. The existing pumps may contain either long driveshafts or the motor may set directly on top of the pump. Both of these older applications can be a pain. Driveshaft units tend to be a maintenance nightmare (carrier bearings, U-joints, and the like). The ones with motors directly on top are susceptible to failure due to flooding. With drypit submersibles, there is no need for driveshafts, and if the station floods, the pumps keep on working (see Photos 11.1 and 11.2). In essence, regardless of what the application is, the same model of pump may work in all three stations.

Even if it is not practical to use the same model, stay with the same manufacturer. This will give you single source availability of original manufacturer parts. Also, one set of necessary spare parts (mechanical seals, packing, o-rings, bearings, etc.) will fit numerous stations. This can save a lot of money.

In the past, we have received quotes to replace old drywell/wetwell combination pump stations with a completely new submersible station. For the size of our stations, the quotes were approximately $100,000.00, compared with the $25,000.00 we spent to replace pumps, all controls, float switches, valves and checkvalves, necessary piping, and even installation. We utilize the existing structures, but theoretically have a new station for approximately one fourth the cost.

There are two major things to watch for in drypit submersible pumps. The first is, specify Class F Insulation. This is important because the pumps will be running at higher temperatures unsubmerged. If they are not rated Class F, they are subject to premature motor failure due to heat. Also, a good drypit submersible pump mechanical seal chamber will use oil for cooling and lubricating. Good quality pump manufacturers

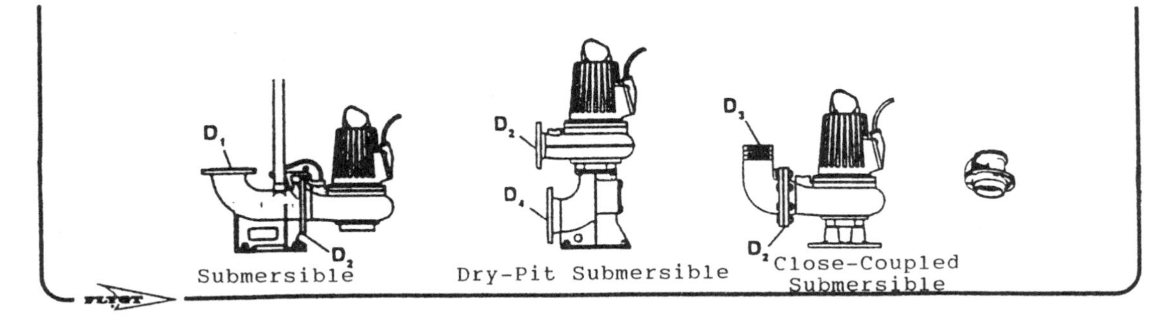

Figure 11.1. Submersible pump types. (Reprinted with permission from ITT Flygt Corporation.)

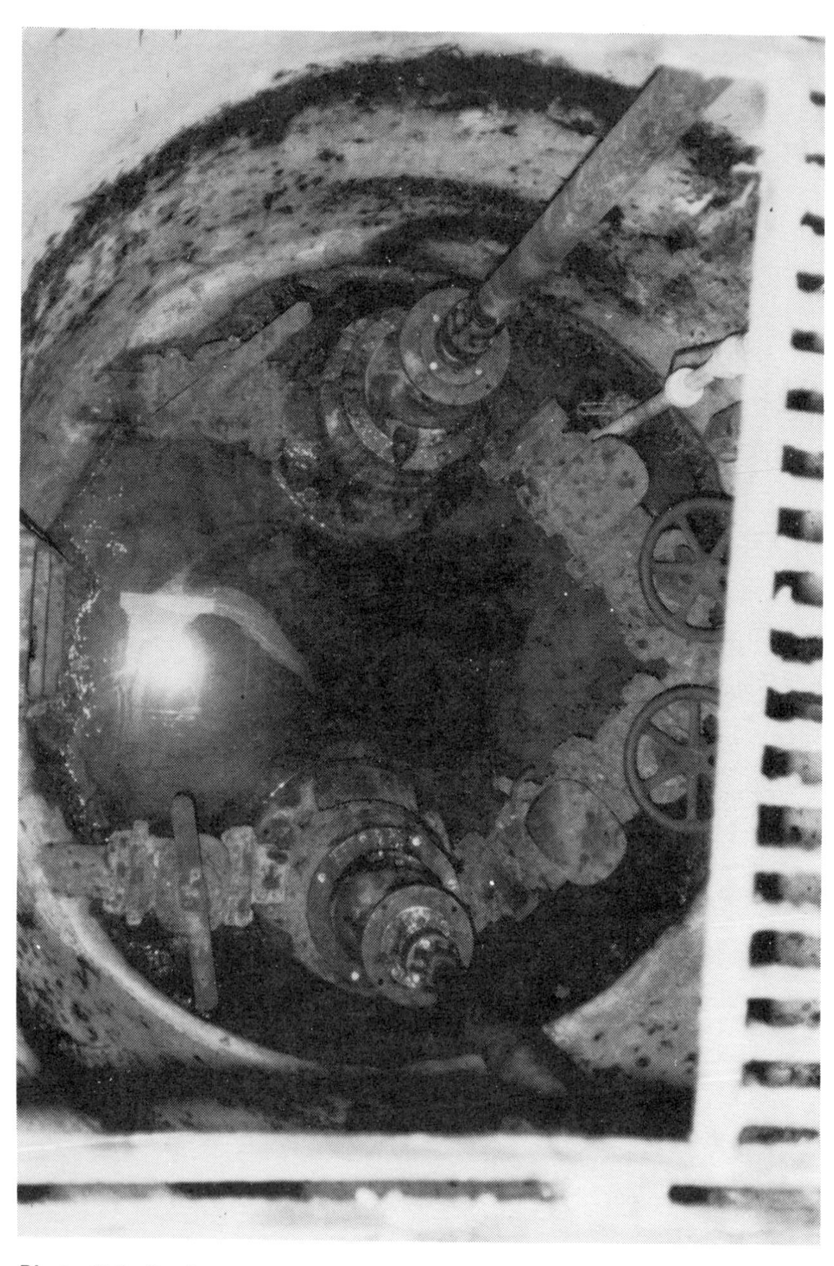

Photo 11.1. Duplex sewage pump station before retrofit of drypit submersible pumps.

Photo 11.2. Same station after dry-pit pump retrofits.

Photo 11.3. Basic control panel used in lift station retrofits.

will recommend changing seal oil at least once a year. This is important as well, considering all submersible and drypit submersible pump seals leak nominal amounts of water. The oil will break down over time and should be changed at least once a year. Also, if you notice a milky color to the oil while changing, the seal either has or is close to failing.

Pump control panels is another of my pet peeves. My motto is "keep it simple, yet effective." For instance, remember when on your most recent upgrade there were a couple of lift station installations included. Upon looking in the panel for the first time it looked like a roadmap of Los Angeles. Wow! And remember the first time the station failed and an electrician was called to repair it and even he couldn't figure it out. There were probably fail-safe circuits on the fail-safe circuits, control systems were layered where each one affected the other, and so on and so forth. There were so many different relays, phase monitors, switches, timers, telephone dialers, etc. that it would cost a mint to store the spare parts. This tends to make a person feel helpless. The way I get around this is to "specify only what is necessary on the system." Typically, this is lightning protection, phase monitors, high level alarms, motor overload protection, and seal failure alarms. If the station was in an area that needed a telephone dialer, then it could be added. Regardless, all our new retrofit stations contain the same control panel, manufactured by the same company. Because the control circuits are all powered by 120 volt, we can stock one set of spare parts for maybe four or five stations. Not only that, but my operators are very familiar with the panels (Photo 11.3) and can troubleshoot any problem very quickly.

Now back to the "high tech control panel" I was speaking of earlier. Generally, the reason for all that excess wiring and controls is it's a way to drive up the project cost. Not always, but more often than not, engineers are compensated based on a percentage of total project cost. I am not a rocket scientist, but I can figure this one out.

CHAPTER 12

Contract Operations

Not many years ago these terms were rarely used. It seems as though recently I cannot pick up an industry magazine, go to a conference, seminar, etc., without hearing the terms *contract operations* or *privatization*. Because they have both been confused with each other through the years, I will attempt to explain what each is about. The term *privatization* refers to a scenario in which a private company actually purchases the treatment works from the municipality. Because this was a very valuable option before 1985, a few municipalities and private companies took advantage of the "tax-exempt laws." Many of these were initiated as design/build/operate contracts that allowed a city to enter into a public/private contract with a firm to provide a turnkey result. Typically, the desired result was to upgrade their existing facilities (sometimes build new facilities), direct all construction, and ultimately operate and maintain the facility for a specified time period (typically, between 5 and 20 years). The main reason was that everything could be put on a fast track for completion due to more flexibility by the private companies.

The most common form of private/public partnerships is contract operations. In this case, the city retains complete ownership of the facilities, and the private company is the manager of its assets.

City governments are realizing it is necessary to operate "lean and mean" to survive. Earlier in the book I spoke of the dwindling federal and state funds that used to be available to municipalities. Because those are becoming fewer and farther between, federal and state unfunded mandates are still being implemented, and Clean Water Act regulations continue to become more stringent, cities are having to do more with less. Of course, you could raise your user rates to compensate for the increased costs, but that is not too popular among sewer users. So the only thing left to do is to cut costs or lose money.

What private operations and maintenance firms are doing is giving entities the option of "contracting out" their operations for a fixed annual cost, they guarantee compliance with all regulatory standards, they take care of any personnel problems associated with the operations, and apply good, solid management practices. In a case in which a city believes they have no other options available due to poor performance from employees, skyrocketing costs to operate, bad public image, etc., contracting with one of these firms would be an option. I know there are cities who employ very good wastewater professionals who want and try to run efficient operations. There are some of those who do not receive the needed support from their boards to get the necessary equipment, resources, manpower, etc. to allow them to operate in an efficient manner. I am a firm believer that many of these problems are a direct result of communication barriers. Because many communities have part-time aldermen, mayors, trustees, etc., they cannot find the time to be involved in the day-to-day operations of their departments. Many times they do not know of problems until they have grown to astronomical proportions.

Communication from both city officials and city employees is vital. If that communication cannot be reached, privatization is an option. The private firms can take the task of upgrading and/or retrofitting equipment and turn it into an art. They typically have engineering personnel on staff or retainer. If they are a very large company at all, they will have experience with similar equipment and associated problems from other facilities they operate. Once again, they know and have all the resources at their fingertips. There may be WWTP operators, superintendents, managers, etc. out there who cannot run an efficient operation working for a municipal government but could shine if working for a private company. It is those specific individuals who have the ability to operate a cost-effective and efficient operation as a city operated entity if given the chance.

One of the biggest lessons I learned while working for a private operation and management firm was to forget all my paradigms. A paradigm is a perceived notion of how someone or something is never going to change. It has always been this way, and that's the way it will stay. This is a mindset of many municipal operators I have met through the years. I never used to think much about it when I would hear one of them say "yeah, we could do a lot more if it wasn't for our mayor" or "John Doe, alderman, is always on my back." Now when I hear that mentality, I instantly think, they either have very little communication between each other or a lack of trust. If a good effort is made by both sides, these barriers can be broken down. The easiest way to do this is a results-oriented operation. If you continually focus on your goals and constantly strive to achieve them, good things will happen.

CHAPTER 13

Conclusion

Keep in mind it took 6 days to create the heavens and earth. Also, keep in mind that nobody has all the answers. It is clearly evident that information is power, and the ability to acquire information and use it properly is an art within itself. Through the years I have found that not always, but typically, my original "gut feelings" about how to address issues were the best. In essence, don't sell yourself short, and never sacrifice your integrity for the good of someone else's interests. If you believe a problem requires a more detailed remedy compared with a quick "band-aid" fix, then stick with your guns and justify the needs till you are blue in the face.

Because this text has been written with operators in mind, I felt it was necessary to touch on not only details of managing a retrofit project but also provide ideas and suggestions for running a professional operation. Everyone has their own ways of handling problems and issues. However, the goal of running a cost-effective and safe operation, producing good effluent quality, and offering a professional service to all tax-paying citizens is the ultimate goal. Due to the fact that the wastewater treatment or sewer treatment field has been labeled as a undesirable task by those who are not aware of the significance or challenges of the profession, I think it is easy for operators to fall into the trap of feeling the same way. Be proud of your profession and continue to look toward the big picture. The big picture is protecting the environment for generations to come.

APPENDIX

Frequently Asked Questions

Q: When do you know it is time to replace or retrofit a piece of equipment?

A: There are several criteria, but the main reasons could be the following:

- excessive man-hours to maintain
- loss of efficiency beyond an acceptable range
- repair costs greater than 50% of replacement
- unavailability of spare parts, etc.

Q: Is the original manufactured equipment still available?

A: Dependent on age, it may still be available. It is best to check with equipment representatives who either sold the original equipment to your entity or have serviced the equipment in the past. (A good source for manufacturer information is a municipal index from industry magazines, i.e., *American City and County, Public Works, Water Environment & Technology*, etc.) Typically, spare parts will be available for 20 years or more from the manufactured date. If the original equipment is not still available, you may find a comparable unit that will fit in the existing structure.

Q: From beginning to end, how long will the retrofit process take?

A: Judging from past experience, a realistic estimate is about 1 year. Of course, this is dependent on the scope of the project and who is going to install the equipment. Equipment usually takes anywhere from 9 to 16 weeks to manufacture. Keep this in mind. The more players involved in the process, the more time the project will take. For instance, a turnkey project aimed at retrofitting a barscreen in which the seller of the equipment is also the installer, will take much less time than if the seller and the installer are two separate entities. Also, if you are installing the equipment yourself, you can cut time and money from the project.

Q: How do I know if I truly need help administering and managing a project?

A: This again depends on the complexity of the project and past experience one may have in executing the necessary tasks involved. The best way to break into this role is by taking on smaller projects first and then building slowly to larger ones as you feel more comfortable.

Q: What is the difference between a "retrofit" and an "upgrade"?

A: A *retrofit* is defined as the replacement of an existing component with another of comparable or same characteristics. An *upgrade* usually entails enlarging or redesigning not only one but possibly multiple components to have greater hydraulic or organic capacity for the facility.

Q: Why even worry about retrofitting equipment; why not wait until the entire facility is decrepit and then replace it all?

A: That rationale is very hard on budgets and subsequently user rates. If equipment is constantly evaluated and replaced one or two units at a time by priority and necessity, the burden will be much less cumbersome. If a couple replacements each year are inserted in the annual budget, then major renovations may be avoided for some time. However, if your board is looking to increase facility capacity for industrial growth or other reasons, facility upgrades may be the only answer.

Q: How do I know if my crew is capable of performing the upgrade in-house?

A: Retrofits utilizing original manufacturer parts are typically relatively simple. A process of writing down and documenting how everything came apart (unwired, unbolted, location of connections, pipe sizes, etc.) will aid in the reconnection and assembly of the new equipment. Other issues to think about are as follows:

- availability of manpower necessary to allocate to the project
- bypassing plans necessary while the equipment is out of service
- special tools needed for the project (Can they be rented, or do they need to be purchased?)
- special permits necessary for the project
- variance necessary on discharge limits during the project, etc.

Q: How do I sell the project to my council or board members?

A: This can be done by communicating to the board equipment problems and associated pitfalls to an operation. Let's say, for instance, you have an old pump station containing packing gland style sealing systems. The station requires 20 man-hours a week to keep operational by repacking pumps, cleaning the floor, the possibility of the station flooding and ruining motors and controls exists, etc. Then convey the expected positive results (less labor intensive, cleaner, sump pumps will not burn out due to being overworked, etc.).

Q: What types, kinds, and how many spare parts should I keep on hand?

A: Typically, the main spare parts to keep on hand are either normal wear types (belts, hoses, fuses, screens, filters, seals, packing, motor overload heaters, gaskets, bearings, etc.) or items that are necessary to maintain effluent quality but may take weeks to manufacture and ship, e.g., gear reducers, motors, specialized electrical controls and components, impellers, valves, etc. Many equipment operation and maintenance manuals contain a list of "Recommended Spare Parts." These recommendations are usually pretty accurate in identifying inventory needs. If you are located close to a metropolitan area, the parts may be obtained easily by driving into town. If, on the other hand, you are located many miles from large distributors, you may want to plan on having a larger variety of spare parts on hand in case of emergencies.

Q: What are some ways in which to sell the public on proposed projects and what mediums do I use?

A: Integrity and trust are the two most important attributes of any public utility that enable the utility to be successful. Delivering what is promised at a good price is also very important. People typically don't mind paying a little more as long as they feel the money is being well spent. Let's say you are wanting to install a more efficient aeration system. The new system will be so efficient, you will be able to shut off one 50-horsepower blower. Provide the council, newspaper, radio, and possibly TV station with a breakdown of proposed savings. Below is the calculation for this information:

50 hp × .7457 = 37 kW,
37 kW × 24 hours = 888 kW hours,
888 kW hours × .06 cents/kW hour = $53.28/day
$53.28/day × 365 days/year = $19,447.20/year

If the complete retrofit costs $75,000.00, the aeration equipment will pay for itself in 3.9 years.

Q: What brand(s) of process equipment is the best and how do I choose from the selection?

A: The best rationale to use in determining which equipment manufacturer is their track record in similar installations. Large process equipment companies have grown because they have good quality equipment and service to offer. This is not to say newer companies do not offer good quality equipment and service; they may not yet have the exposure of older, more established companies. It is better to go with equipment that has proven itself time and time again, rather than being a "guinea pig" for something new.

Q: How do I communicate positive results to the council and community without the risk of sounding like I am ringing our own bell?

A: Stick with nonsubjective data. If the retrofit, or upgrade is saving thousand's of dollars in electric bills, show a breakdown of the savings. If man-hours to maintain a retrofitted piece of equipment have dropped, list how many man-hours are saved. If effluent quality has improved by a certain percentage or you have returned to compliance due to the retrofit, show graphs and numbers illustrating this.

Index